All readers, irrespective of their academic specialty, will find this book stimulating as it raises critical issues around the need for respect for the earth and the restoration of traditional ecological ethos to live in harmony with Nature. Following the global tragedy caused by COVID-19 and the unpredictable nature of zoonotic pandemics, Ngozi Unuigbe argues for a complement of traditional ecological knowledge and mainstream science in mitigating future pandemics.

The scope and variety of literature explored demonstrates the need for a Multiple Evidence-Based (MEB) approach to identify knowledge gaps and the opportunities for collaborative research, as well as the need for urban planning and adaptive management.

This well-written volume promotes the need for better recognition of environmental ethics and planetary health for the well-being of the entire ecosystem. It is a must-read for everyone interested in the sustainability of natural resources and the empowerment of indigenous communities.

—Doyin Ogunyemi, Public Health Physician,
University of Lagos, Nigeria.

Indigenous wisdom is essential to solving the global ecological crisis. In particular, traditional ecological knowledge can help inspire a new generation of Earth-centred laws, sciences, social norms and ways of living in harmony with Nature. In her book *Traditional Ecological Knowledge and Global Pandemics*, Professor Ngozi Finette Unuigbe takes a critical look at today's most pressing environmental challenges and puts forward a compelling blueprint for how indigenous world views can galvanize a thriving future for the planet.

—Grant Wilson, Executive Director of Earth Law Center and editor of *Earth Law: Emerging Ecocentric Law – A Guide for Practitioners*.

Finette Unuigbe's *Traditional Ecological Knowledge and Global Pandemics* is a clarion call for attending to what is already known but unfortunately often forgotten – the role that nature plays in making the world habitable, hospitable and possible for humans. This book makes an important case in support of ecological diversity and intactness to prevent the incubation and spread of zoonotic and other diseases, including COVID-19. Policymakers and medical health professionals in particular should read it before the sun sets.

—James R. May, Esq. Professor of Law, Widener University Delaware Law School; Representative for Rights of Nature, International Council of Environmental Law.

The book is of an astonishing actuality. It is underpinned by an excellent literature review, which proves biodiversity loss and global health burdens are intrinsically related. COVID-19 and other pandemics of this century have a common origin: the destruction of nature. We are nature and the suffering of mother earth is our pain.

The greatest achievement of this book is nonetheless the solid description of how traditional ecological knowledge and mainstream scientific knowledge can work together and find a 'peaceful coexistence' between humans and non-human beings. Starting from important evidence that challenges the western common sense, the book proves that the way indigenous peoples manage their living place is the ground of keeping nature balanced and flourishing.

Finally, it invites us to accept the challenge of having the best of the two worlds, combining these two knowledge systems for a co-production and an intercultural brokerage. We are convinced that there is no other way to survive the pandemics and live in harmony with nature on this Earth – Gaia. Not only the reading of this book is necessary, but above all putting its content into practice is urgent.

—Christianne Derani, Professor in International Economic and Environmental Law, Federal University of Central Catarina, Brazil. Formerly, Instructor to Indigenous Peoples on Biodiversity Conservation.

Traditional Ecological Knowledge and Global pandemics

This book demonstrates the importance and potential role of traditional ecological knowledge in foreseeing and curbing future global pandemics.

The reduction of species diversity has increased the risk of global pandemics and it is therefore not only imperative to articulate and disseminate knowledge on the linkages between human activities and the transmission of viruses to humans, but also to create policy pathways for operationalizing that knowledge to help solve future problems. Although this book has been prompted by the COVID-19 pandemic, it lays a policy foundation for the effective management or possible prevention of similar pandemics in the future. One effective way of establishing this linkage with a view to promoting planet health is by understanding the traditional ecological knowledge of indigenous peoples with a view to demonstrating the significant impact it has on keeping nature intact. This book argues for the deployment of traditional ecological knowledge for land use management in the preservation of biodiversity as a means for effectively managing the transmission of viruses from animals to humans and ensuring planetary health. The book is not projecting traditional ecological knowledge as a panacea to pandemics but rather accentuating its critical role in the effective mitigation of future pandemics.

This book will be of great interest to indigenous peoples, policy makers, students and scholars of traditional ecological knowledge, indigenous studies, animal ecology, environmental ethics and environmental studies more broadly.

Ngozi Finette Unuigbe is a researcher in international environmental law, policy and ethics whose research engages the stewardship ethic of indigenous knowledge, within the context of biodiversity preservation. She is currently a Professor at the University of Benin, Nigeria.

Routledge Focus on Environment and Sustainability

Reframing Energy Access
Insights from The Gambia
Anne Schiffer

Climate and Energy Politics in Poland
Debating Carbon Dioxide and Shale Gas
Aleksandra Lis

Sustainable Community Movement Organizations
Solidarity Economies and Rhizomatic Practices
Edited by Francesca Forno and Richard R. Weiner

Climate Change Ethics for an Endangered World
Thom Brooks

The Emerging Global Consensus on Climate Change and Human Mobility
Mostafa M Naser

Traditional Ecological Knowledge in Georgia
A Short History of the Caucasus
Zaal Kikvidze

Traditional Ecological Knowledge and Global Pandemics
Biodiversity and Planetary Health Beyond Covid-19
Ngozi Finette Unuigbe

Climate Diplomacy and Emerging Economies
India as a Case Study
Dhanasree Jayaram

Linking the European Union Emissions Trading System
Political Drivers and Barriers
Charlotte Unger

For more information about this series, please visit: www.routledge.com/Routledge-Focus-on-Environment-and-Sustainability/book-series/RFES

Traditional Ecological Knowledge and Global Pandemics
Biodiversity and Planetary Health Beyond Covid-19

Ngozi Finette Unuigbe

LONDON AND NEW YORK

First published 2021
by Routledge
2 Park Square, Milton Park, Abingdon, Oxon OX14 4RN

and by Routledge
52 Vanderbilt Avenue, New York, NY 10017

Routledge is an imprint of the Taylor & Francis Group, an informa business

© 2021 Ngozi Finette Unuigbe

The right of Ngozi Finette Unuigbe to be identified as author of this work has been asserted by them in accordance with sections 77 and 78 of the Copyright, Designs and Patents Act 1988.

All rights reserved. No part of this book may be reprinted or reproduced or utilised in any form or by any electronic, mechanical, or other means, now known or hereafter invented, including photocopying and recording, or in any information storage or retrieval system, without permission in writing from the publishers.

Trademark notice: Product or corporate names may be trademarks or registered trademarks, and are used only for identification and explanation without intent to infringe.

British Library Cataloguing-in-Publication Data
A catalogue record for this book is available from the British Library

Library of Congress Cataloging-in-Publication Data
A catalog record has been requested for this book

ISBN: 978-0-367-69294-0 (hbk)
ISBN: 978-1-003-14128-0 (ebk)

Typeset in Times
by Deanta Global Publishing Services, Chennai, India

I specially thank Moyosore Olopade and Osaheni Benedict Ezomo for their efforts in copyediting this work; and of course, my family and many friends…my achievement is yours, too.

Contents

1 Pandemics and the environment 1

2 Why traditional ecological knowledge matters 17

3 Complementarity of traditional ecological knowledge and mainstream science 27

4 Restoring land, community and health 36

5 Synergizing TEK and mainstream science to promote planetary health 53

Index 79

1 Pandemics and the environment

> We have troubled the habitats of viruses and caused them to relocate to…us.

Introduction

The right to health is well established as a fundamental right of every human being (WHO Constitution 1946). Biodiversity is at the heart of the intricate web of life on Earth and the processes essential to its survival. Our planet's biological resources are not only shaped by natural evolutionary processes but are also increasingly transformed by anthropogenic activities, population pressures and globalizing tendencies. When human activities threaten these resources, or the complex ecosystems of which they are a part, they pose potential risks to millions of people whose livelihoods, health and well-being are sustained by them.

The increasingly complex global health challenges that we face, including poverty, malnutrition, infectious diseases and the growing burden of non-communicable diseases (NCDs), are more intimately tied than ever to the complex interactions between ecosystems, people and socioeconomic processes. These considerations are also at the heart of the post-2015 Development Agenda and the Sustainable Development Goals (SDGs). The dual challenges of biodiversity loss and rising global health burdens are not only multifaceted and complex; they also transcend sectoral, disciplinary and cultural boundaries, and demand far-reaching, coherent and collaborative solutions (UNEP 2015).

Human mistreatment of the natural environment has turned out to have distinctly painful boomerang effects. The ongoing destruction of the Amazon rain forest, for example, is altering Brazil's climate, raising temperatures and reducing rainfall in significant ways, with painful consequences for local farmers and even more distant urban dwellers. (And the release of vast quantities of carbon dioxide, from increasingly massive forest fires, will only increase the pace of climate change globally.) Similarly, the technique of hydraulic fracking, used to extract oil and natural gas trapped in underground shale deposits, can trigger earthquakes that damage aboveground structures and endanger human life.

The sorts of large environmental changes we are seeing today, including climate and land use change, have a high potential to lead to changes in health outcomes, including the transmission of infectious diseases. Studies are beginning to show how environmental destruction could lead to a greater spread of deadly human diseases via animals and other organisms, with serious consequences for public health.

It is therefore correct to say Nature is striking back. While many climate effects like prolonged heatwaves will become more pronounced over time, other effects, it is now believed, will occur suddenly with little warning and could result in large-scale disruptions in human life.

Environmental tipping points

Until now, the tipping points of greatest concern to scientists have been the rapid melting of the Greenland and West Antarctic ice sheets. Those two massive reservoirs of ice contain the equivalent of hundreds of thousands of square miles of water. Should they melt ever more quickly with all that water flowing into neighbouring oceans, a sea level rise of 20 ft or more can be expected, inundating many of the world's most populous coastal cities and forcing billions of people to relocate (Pörtner et al. 2019). In its 2014 study, the Intergovernmental Panel on Climate Change (IPCC) predicted that this might occur over several centuries, at least offering plenty of time for humans to adapt, but more recent research indicates that those two ice sheets are melting far more rapidly than previously believed – and so a sharp increase in sea levels can be expected well before the end of this century with catastrophic consequences for coastal communities (IPCC's Fifth Assessment Report 2014).

The IPCC also identified two other possible tipping points with potentially far-reaching consequences: the die-off of the Amazon rain forest and the melting of the Arctic ice cap. Both are already under way, reducing the survival prospects of flora and fauna in their respective habitats. As these processes gain momentum, entire ecosystems are likely to be obliterated and many species killed off, with drastic consequences for the humans who rely on them in so many ways (from food to pollination chains) for their survival. But as is always the case in such transformations, other species – perhaps insects and microorganisms highly dangerous to humans – could occupy those spaces emptied by extinction (Ibid).

Back in 2014, the IPCC did not identify human pandemics among potential climate-induced tipping points, but it did provide plenty of evidence that climate change would increase the risk of such catastrophes. This is true for several reasons. First, warmer temperatures and more moisture are conducive to the accelerated reproduction of mosquitoes, including those carrying malaria, the Zika virus and other highly infectious diseases. Such conditions were once largely confined to the tropics, but as a result of global warming, formerly temperate areas are now experiencing more tropical conditions, resulting in the territorial expansion of mosquito breeding grounds. Accordingly, malaria and Zika are on the rise in areas that never previously experienced such diseases. Similarly, dengue fever, a mosquito-borne viral disease that infects millions of people every year, is spreading especially quickly due to rising world temperatures (Caminade et al. 2019).

Combined with mechanized agriculture and deforestation, climate change is also undermining subsistence farming and indigenous lifestyles in many parts of the world, driving millions of impoverished people to already crowded urban centres, where health facilities are often overburdened and the risk of contagion ever greater (Benett 2017; Hansungule & Jegede 2014). Virtually all the projected growth in populations will occur in urban agglomerations. Adequate sanitation is lacking in many of these cities, particularly in the densely populated shantytowns that often surround them (Chan 2017). About 150 million people currently live in cities affected by chronic water shortages, and by 2050, unless there are rapid improvements in urban environments, the number will rise to almost a billion (Majumder 2015).

Such newly settled urban dwellers often retain strong ties to family members still in the countryside who, in turn, may come in contact with wild animals carrying deadly viruses. This appears to have been the origin of the West African Ebola epidemic of 2014–2016, which affected tens of thousands of people in Guinea, Liberia and Sierra Leone. Scientists believe that the Ebola virus originated in bats and was then transmitted to gorillas and other wild animals that co-exist with people living on the fringes of tropical forests (Carl-Johan 2015). Somehow, a human or humans contracted the disease from exposure to such creatures and then transmitted it to visitors from the city who, upon their return, infected many others. Other diseases that have crossed into humans include Lassa fever, which was first identified in 1969 in Nigeria; Nipah from Malaysia and SARS from China, which killed more than 700 people and travelled to 30 countries in 2002–03. Some, like Zika and West Nile virus, which emerged in Africa, have mutated and become established on other continents (Ibid).

The coronavirus disease that emerged in China in December 2019 (COVID-19) appears to have had somewhat similar origins. In recent years, hundreds of millions of once impoverished rural families moved to burgeoning industrial cities in central and coastal China, including places like Wuhan. Although modern in so many respects, with its subways, skyscrapers and superhighways, Wuhan also retained vestiges of the countryside, including markets selling wild animals still considered by some inhabitants to be normal parts of their diet. Many of those animals were trucked in from semi-rural areas hosting large numbers of bats, the apparent source of both the coronavirus and the Severe Acute Respiratory Syndrome, or SARS, outbreak of 2013, which also arose in China. Scientific research suggests that breeding grounds for bats, like mosquitoes, are expanding significantly as a result of rising world temperatures (Woo et al. 2006; Woodward 2020).

As exemplified by coronaviruses and influenza viruses, bats and birds are natural reservoirs for providing viral genes during the evolution of new virus species and viruses for interspecies transmission. These warm-blooded vertebrates display high species biodiversity, roosting and migratory behaviour, and a unique adaptive immune system, which are favourable characteristics for asymptomatic shedding, dissemination and mixing of different viruses for the generation of novel mutant, recombinant or reassortant RNA viruses. The increased intrusion of humans into wildlife habitats and overcrowding of different wildlife species in wet markets and farms have also facilitated the interspecies transmission between different animal species (Chan et al. 2013).

Are we therefore facilitating the emergence and spread of deadly pathogens like the Ebola virus, SARS and the coronavirus through deforestation, haphazard urbanization and the ongoing warming of the planet? It may be too early to answer such a question unequivocally, but the evidence is growing that this is the case. If so, actions to mitigate this trend are imperative. In fact, a new discipline, planetary health, is emerging that focuses on the increasingly visible connections among the well-being of humans, other living things and entire ecosystems.

A look at the origins of COVID-19 reveals that other forces may be in play. In the past century we have escalated our demands upon nature, such that today we are losing species[1] at a rate unknown since the dinosaurs, along with half of life on Earth, went extinct 65 million years ago. This rapid dismantling of life on Earth owes primarily to habitat loss, which occurs mostly from growing crops and raising livestock for people. With fewer places to live and fewer food sources to feed on, animals find food and shelter where people are, and that can lead to disease spread (Roman-Palados 2020).

As seen above, a major cause of species loss is climate change, which can also change where animals and plants live and affect where diseases may occur. Historically, we have grown as a species in partnership with the plants and animals we live with. So, when we change the rules of the game by drastically changing the climate and life on Earth, we have to expect that it will affect our health. The dangers posed by fruit bats and mosquitoes are rarely mentioned among the potential impacts of major environmental changes such as deforestation and climate change.

The coronavirus may not, in retrospect, prove to be *the* tipping point that upends human civilization as we know it, but it should serve as a warning that we will experience ever more such events in the future as the world heats up. The only way to avert such a catastrophe and assure ourselves that Nature will not become an avenger is to cease the further desecration of essential ecosystems.

Zoonoses and pandemics

Human-caused global changes, such as deforestation, activities of extractive industries including logging and mining, introduction of invasive species and urban development, are driving infectious disease emergence and spread, as well as biodiversity loss. There are opportunities for preventing infectious diseases and reducing biodiversity loss by addressing their common drivers through a synergistic approach (UNEP 2015).

From the foregoing section, it is evident that whether its origin is from a bat or pangolin, the coronavirus outbreak that has killed tens of thousands and turned the world upside down comes from the animal world. It is human activity that enabled the virus to jump to people, and specialists are warning that if nothing changes many other pandemics of this nature will follow. In a review of 1,407 species of human pathogenic organisms, 816 (58%) were broadly classified as zoonotic (Woolhouse & Gowtage–Sequeria 2005), a term coined by Virchow and defined by the World Health Organization in 1959 to describe 'those diseases and infections (the agents of) which are naturally transmitted between (other) vertebrate animals and man' (Mantovani 2001). The word is based on the Greek words for 'animal' and 'sickness'. They are not new – tuberculosis, rabies, toxoplasmosis, malaria, to name just a few, are all zoonoses. According to the UN Environment Programme (UNEP), 60% of human infectious diseases originate from animals. This figure climbs to 75% for 'emerging' diseases such as Ebola, HIV, avian flu, Zika or SARS – another type of coronavirus. The list goes on (UNEP 2020).

Emerging infectious diseases, most of which are considered zoonotic in origin, continue to exact a significant toll on society. The origins of major human infectious diseases are reviewed and the factors underlying disease emergence explored. Anthropogenic changes (largely in land use and agriculture) are implicated in the apparent increased frequency of emergence and re-emergence of zoonoses in recent decades. Special emphasis is placed on the pathogen with likely the greatest zoonotic potential, influenza virus.

The concept of 'emerging infectious diseases' has changed from a mere curiosity in the field of medicine to an entire discipline that has been gaining prominence (Spec et al. 2020). In recent decades, previously unknown diseases have surfaced at a pace unheard of in the recorded annals of medicine – more than 30 newly identified human pathogens in 30 years, most of them newly discovered zoonotic viruses. The exact proportion of emerging human diseases that have arisen from non-human animals is unknown. However, according to the Woolhouse and Gowtage-Sequeria review, of the 177 of 1,407 human pathogens identified as 'emerging', 130 (73%) were zoonotic (Woolhouse & Gowtage–Sequeria 2005; Spec et al. 2020). The landmark U.S. Institute of Medicine (IOM) report *Emerging Infections* concluded that 'the significance of zoonoses in the emergence of human infections cannot be overstated' (Greger 2007).

The first surge in species-barrier breaches likely occurred with the clustering of zoonotic vectors accompanying the domestication of animals – but that was 10,000 years ago (Woolhouse & Gowtage–Sequeria 2005). What new changes are taking place at the human/animal interface that may be responsible for this resurgence of zoonotic disease in recent decades?

In 2004, a joint consultation on emerging zoonoses was convened by the World Health Organization, the Food and Agriculture Organization of the United Nations and the World Organization for Animal Health to elucidate the major drivers of zoonotic disease emergence (Greger 2007). A common theme of primary risk factors for the emergence and spread of emerging zoonoses was the increasing demand for animal protein, associated with the expansion and intensification of animal agriculture, long-distance live animal transport, live animal markets, bushmeat consumption and habitat destruction (Greger 2007; Daszak 2012; Emerging Zoonoses 2020).

Livestock Revolution

Driven by the population explosion, urbanization and increasing incomes, the per-capita consumption of meat, eggs and dairy products has dramatically expanded in the developing world, leading to what has been termed the 'Livestock Revolution' beginning in the 1970s, akin to the 1960s Green Revolution in cereal grain production. From around 1980 to date, world meat production has more than doubled and this trend will increase to the growing demand (FAO 2018).

Animal agriculture worldwide is increasingly moving from the relatively low efficiency, family-centred, low-input model to intensive systems, which are loosely defined as the production of large numbers of genotypically similar animals often under concentrated confinement with rapid population turnover. Traditional systems are being replaced by intensive systems at a rate of more than 4% per year, particularly in Asia, Africa and South America. While heralded for its efficiency and productive capacity, this industrial model has raised sustainability concerns regarding the waste absorption and feed supply capacity of available land, as well as trepidation over associated zoonotic risks. This intensive industrialization of animal agriculture may represent the most profound alteration of the animal–human relationship since domestication. Given the emergence of some of humanity's most important diseases in the Neolithic era, there is concern that new threats may arise from the broadly significant changes currently taking place in global livestock production (Jones 2013).

Changes in land use or agriculture have been identified as the main driver of the appearance of emerging infectious diseases. Unnaturally high concentrations of animals confined indoors in a limited airspace and producing significant quantities of waste may allow for the rapid selection, amplification and dissemination of zoonotic pathogens. These increasingly greater numbers of animals into increasingly smaller spaces has been identified as a critical factor in the spread of disease (Vanwambeke et al. 2019).

The industrialization of animal production may lead not only to greater animal-to-animal contact, but increasing animal-to-human contact, particularly when production facilities border urban areas. Though land pressures have tended to push crops and extensive animal systems away from the developing world's growing megacities, intensive livestock operations are moving closer to major urban areas in countries such as Bangladesh. This nexus, described as the 'peri-urbanization' of industrial animal agriculture, may provide 'flash points' for the source of new diseases (Ibid).

Livestock transport

Over the last century, there has been a shift away from livestock production as a highly localized enterprise, where animals were typically born, fattened and slaughtered in the same region. The number of live animals traded for food quintupled in the 1990s, with huge numbers moved across borders. Long-distance live animal transport may make countries more vulnerable to acts of bioterrorism with zoonotic agents, a risk thought to be amplified by the concentrated and intensive nature of contemporary industrial farming practices. The transport and crowding of animals from different herds or flocks in poorly ventilated and stressful environments for long periods has been considered ideally suited for spreading disease (Spickler 2015).

Bushmeat consumption

Humans and our ancestors have likely consumed bushmeat, wild animals killed for food, for millions of years. During the 20th century, however, commercial hunting using firearms and wire snares to supply logging and oil exploration operation concessions along new roadway networks has dramatically increased the catch in Central African forest. Annually, it is estimated that 579 million wild animals are caught and consumed in the Congo basin, equalling 4.5 million tons of bushmeat, with the addition of a possible 5 million tons of wild mammalian meat from the Amazon basin (Enuoh & Bisong 2014).

Tropical lowland forest habitat contains the world's greatest terrestrial biodiversity and may therefore harbour a reservoir of zoonotic pathogens. Logging in Central Africa, for example, generally involves selective extraction of high-value timber species rather than clear-cutting, which may maintain this higher natural density of potential hosts (Losh 2020). It is estimated that the wildlife trade in general generates in excess of one billion direct and indirect contacts between humans and domesticated animals annually. The broad range of tissue and fluid exposures associated with the bushmeat industry's hunting and butchering may make these wildlife interactions especially risky (Enuoh & Bisong 2014).

Habitat destruction

In 1933, wildlife ecology pioneer Aldo Leopold reportedly wrote that the real determinants of disease mortality are the environment and the population, both of which he felt were being 'doctored daily, for better or for worse, by gun and axe, and by fire and plow' (Friend et al. 2001). Since Leopold penned those lines, more than half of the earth's tropical forests have been cleared. Based on satellite imaging, global tropical deforestation continues at an annual loss of up to 2–3%, with the net rate of tropical forest clearing increasing roughly 10% from the 1980s to 1990s (Helmholtz Centre for Environmental Research 2018).

Grazing animals for human consumption demands an estimated 0.21 hectares per global capita, comparable with the 0.22 hectares attributed to timber harvesting. One hundred thousand years ago, likely all members of the human race lived in eastern Africa. The world's population grew from fewer than 100 million people 3,000 years ago to approximately one billion at the turn of the 20th century, more than six billion by the end of the century, with 2050 projections approaching ten billion. Two-thirds of this population increase occurred within the last 50 years, leading to major ecological changes and wildlife habitat reduction implicated in zoonotic disease emergence (Bannister-Tyrell et al. 2015).

Rajan Patil, associate professor of epidemiology at SRM University, Chennai, says growing proximity between human settlements and wildlife is increasing the rate of disease transmission between domestic animals and wildlife. He reiterates that irrespective of whether humans are going into forest areas or animals are coming to human settlements due to deforestation, viruses are being exchanged (Padma 2020).

He asserts that usually in undisturbed habitats, viruses keep circulating in mild forms in animals. It is when this equilibrium is disturbed and they come in contact with humans, some cross the species barrier due to a mutation, and human infections start taking place. He cited the example of swine flu (H1N1) virus, which remains a very mild infection in pigs, but becomes 'deadly' when it mutates and crosses over to humans.

Other examples cited by Patil are Lyme disease, transmitted by ticks through white-footed mice; West Nile Disease, a mosquito-borne disease whose primary reservoir is wild birds; and an outbreak of anthrax in Chhattisgarh state which he attributes to the loss of biodiversity.

A key area of concern is deforestation to make way for agriculture and intensive livestock farming. Domesticated animals are often a 'bridge' between pathogens from the wild and humans. The widespread use of antibiotics in the livestock industry has also led to bacterial pathogens building up immunity to front-line drugs. Urbanization and habitat fragmentation are also highly disruptive of the balance between species, while global warming can push disease-carrying animals into new territory (Padma 2020).

Experts are increasingly linking deforestation with recent outbreaks. Researchers working on the Ebola virus have shown how the index cases of about ten Ebola epidemics in Africa over the last two decades occurred in regions affected by deforestation and forest fragmentation. According to them, 'the spillover from wildlife or reservoir species into humans is favoured by land use change such as deforestation and forest fragmentation for a variety of reasons'. One is that as humans encroach wildlife habitats, there are increased chances of human contact with infected species. A second is that natural habitat destruction alters animal community dynamics, and sometimes increases the numbers of some 'generalist' pathogens that can reside in a range of hosts, and reduces the numbers of 'specialist' species that thrive in limited hosts in the previous 'undisturbed' conditions (Ibid).

Thomas Gillespie, associate professor at the department of environmental sciences at Emory University, United States, says that when extractive industries, such as logging, oil exploration and mining, are implemented in largely uninhabited wilderness areas, they provide the opportunity for human exposure to novel pathogens.

He claims that while scientists do not have the necessary data yet on where exactly the dislodged bats went or how they adapted, 'but there are plausible linkages'. 'Whenever you have novel interactions with a diverse range of species in one place – whether that's in a natural environment like a tropical forest or in an artificially created environment like a wet market – you can have spillover events', says Gillespie (Ibid). He argues that the wet markets really represent the minority of opportunities for spillover to occur. Close to a third of diseases that emerge are linked to large-scale land use change like deforestation and well over half of diseases that emerge are coming from wildlife in forests – including such well-known examples as HIV and Ebola (Ibid).

Beyond the current outbreak of coronavirus, it has been estimated that zoonoses kill some 700,000 people a year (Jay-Russel & Doyle 2016). According to researchers, this coronavirus outbreak may just be the tip of the iceberg. Increased trends in land use change, combined with increased trends in trade and global travels, are expected to increase the frequency of pandemics in the future (Woolhouse & Gowtage–Sequeria 2005; Spec et al. 2020). Transformative change is needed in order to find a solution to this global tragedy – a systemic response. Environmental policies that promote sustainable land use planning, reduced deforestation and biodiversity protection provide ancillary benefits by reducing the types of wildlife contact that can lead to disease emergence (Estrada-Pena et al. 2014).

Towards a planetary health policy

From the foregoing, the following deductions can be made about zoonotic disease transmission:

- nearly two-thirds of human infectious diseases arise from pathogens shared with wild or domestic animals;
- endemic and enzootic zoonoses cause about a billion cases of illness in people and millions of deaths every year, and emerging zoonoses are a rising threat to global health;
- ecological and evolutionary perspectives can provide valuable insights into pathogen ecology and can inform zoonotic disease-control programmes;
- anthropogenic practices, such as changes in land use and extractive industry actions, animal production systems, and widespread antimicrobial applications affect zoonotic disease transmission;
- risks are not limited to low-income countries; as global trade and travel expands, zoonoses are increasingly posing health concerns for the global medical community;
- ecological, evolutionary, social, economic and epidemiological mechanisms affecting zoonoses' persistence and emergence are not well understood; such information could inform evidence-based policies, practices, and targeted zoonotic disease surveillance, and prevention and control efforts;

- multisectoral collaboration among clinicians, public health scientists, ecologists and disease ecologists, veterinarians, economists and others is necessary for effective management of the causes and prevention of zoonotic diseases.

Overall, zoonotic and food-borne diseases have an effect across society, with everyone sharing some burden. As in the period of the 19th century when many veterinary and human health systems were initiated, there is a need to re-examine how existing systems are structured, resourced and managed to create synergies between animal and human health and in the process reduce the effect of zoonotic disease burdens. This process requires an evidence base to effectively, systematically and strategically inform policy developments.

Advances in public health, veterinary and human medicine offer benefits, but only respond to part of the struggle that we face. Many stakeholders have important roles in researching, planning and implementing efforts to prevent, contain and mitigate emerging infectious diseases at levels that stretch from the community to worldwide. These people work in many sectors and are involved in responding to the challenge as a primary aim, a secondary aim or as a by-product. They include, for example, wildlife management, farming and agriculture, veterinary medicine, the pharmaceutical industry, human public health, non-governmental organizations, the donor community, ministerial policymakers and UN agencies, to name but a few. Moreover, many academic disciplines contribute essential knowledge (e.g., climatologists, plant scientists, molecular biologists, economists, political scientists). It is imperative to have an overarching or grand narrative to link all the sectors and stakeholders, and we need this narrative to have a framework.[2]

This broad view is essential for the successful development of policies and practices that reduce the probability of future zoonotic emergence, targeted surveillance and strategic prevention, *and engagement of partners outside the medical community to help improve health outcomes and reduce disease threats* (Kareh et al. 2012).[3] It cannot therefore be overemphasized that beyond the essential response to each pandemic, we must re-think our model and gravitate towards one that is more in harmony with Nature.

Conclusion: future directions

Suppose the interpretation of the COVID-19 pandemic in the foregoing sections is correct; suppose that the coronavirus is Nature's warning, her way of telling us that we have gone too far and must alter our behaviour lest we risk further contamination. What then? What humanity may need to do is institute a new policy of 'peaceful coexistence' with Nature. This approach would legitimize the continued presence of large numbers of humans on the planet but require that they respect certain limits in their interactions with its ecosphere. We humans could use our talents and technologies to improve life in areas we have long occupied, but infringement elsewhere would be heavily restricted. Natural disasters – floods, volcanoes, earthquakes and the like – would, of course, still occur, but not at a rate exceeding what we experienced in the pre-industrial past.

Implementation of such a strategy would require, at the very least, mitigating climate change-induced risks as swiftly as possible through the rapid and thorough elimination of human-induced carbon emissions – something that has, in fact, happened in at least a modest way due to measures to contain the spread of COVID-19. Essentially, deforestation would also have to be reduced to the barest minimum and the world's remaining wilderness areas preserved; also, further unsustainable agricultural practices would have to be discouraged. In addition, any further despoliation of the oceans would have to be stopped, including the dumping of wastes, plastics, engine fuel and runoff pesticides.

Change must come from both rich and poor societies. Demand for wood, minerals and resources from the Global North leads to the degraded landscapes and ecological disruption that drives disease. The risks are greater now. They were always present and have been there for generations. It is our interactions with that risk which must be changed. We are in an era now of chronic emergency. Diseases are more likely to travel further and faster than before, which means we must be faster in our responses. It needs investments, change in human behaviour, and it means we must listen to (involve) people at community levels.

Also, we may rethink urban infrastructure, particularly within low-income and informal settlements. Short-term efforts are focused on containing the spread of infection. The longer-term – given that new infectious diseases will likely continue to spread rapidly into and within cities – calls for an overhaul of current approaches to urban planning and development. The bottom line is to be prepared. We cannot predict where the next pandemic will come from, so we need mitigation plans to take into account the worst possible scenarios. The only certain thing is that the next one will certainly come. As seen above, hunting, farming and the global move of people to cities has led to massive declines in biodiversity and increased the risk of dangerous viruses like COVID-19 spilling over from animals to humans.

Domesticated animals like cattle, sheep, dogs and goats shared the highest number of viruses with humans, with eight times more animal-borne viruses than wild mammal species. Wild animals that have adapted well to human-dominated environments also share more viruses with people.

Rodents, bats and primates – which often live among people, and close to houses and farms – together were implicated as hosts for nearly 75% of all viruses. Bats alone have been linked to diseases like SARS, Nipah, Marburg and Ebola (Georgiou 2020).

The spillover risk is highest from threatened and endangered wild animals whose populations had declined largely due to hunting, the wildlife trade and loss of habitat. Human encroachment into biodiverse areas increases the risk of spillover of novel infectious diseases by enabling new contacts between humans and wildlife. Species in the primate and bat orders are significantly more likely to harbour zoonotic viruses compared to all other orders. Spillover of viruses from animals are a direct result of our actions involving wildlife and their habitat.

The consequence is they are sharing their viruses with humans. These actions simultaneously threaten species survival and increase the risk of spillover. In an unfortunate convergence of many factors, this brings about the kind of challenge humanity is in. There is the need to rethink how we interact with wildlife and the activities that bring humans and wildlife together. We obviously do not want pandemics of this scale. We need to find ways to co-exist safely with wildlife, as they have no shortages of viruses to transmit to humans.

Separately, more than 200 of the world's wildlife groups have written to the World Health Organization (WHO) calling on it to recommend to countries a highly precautionary approach to the multi-billion-dollar wildlife trade, and a permanent ban on all live wildlife markets and the use of wildlife in traditional medicine. The COVID-19 pandemic, says the letter, is believed to have originated at wildlife markets in China, and to have been transmitted to humans as a result of the close proximity between wildlife and people.

The groups, which include the International Fund for Animal Welfare, the Zoological Society of London and People for the Ethical Treatment of Animals (PETA), say a ban on wildlife markets globally will help prevent the spread of disease, and address one of the major drivers of species extinction. This decisive action, well within the WHO's mandate, would be an impactful first step in adopting a highly precautionary approach to wildlife trade that poses a risk to human health. The organizations argue that zoonotic diseases are responsible for over two billion cases of human illness and over two million human deaths each year, including from Ebola, Mers, HIV, bovine tuberculosis, rabies and leptospirosis.

To drive the foregoing recommendations, longer-term strategies should be deployed globally. One of such is to focus on behavioural change through a gradual change of ethics. This interplay between human activity and planetary behaviour has led some analysts to rethink our relationship with the natural world. They have reconceptualized the earth as a complex matrix of living and inorganic systems, all (under normal conditions) interacting to maintain a stable balance. When one component of the larger matrix is damaged or destroyed, the others respond in their unique ways in attempting to restore the natural order of things. This notion has often been described as 'the Gaia Hypothesis' after the ancient Greek goddess Gaia, the ancestral mother of all life.

Notes

1. *Species* refers to the number and types of plants and animals that exist in a particular area or in the world generally. In ecology, a *habitat* is the type of *natural* environment in which a particular species of organism lives. A species' habitat is those places where the species can find food, shelter, protection and mates for reproduction. It is characterized by both physical and biological features. *Biological diversity (biodiversity)* in an environment comprises numbers of different species of plants and animals. Throughout this book, the term 'biodiversity conservation' or 'species conservation' will be used interchangeably to reflect the focus of the book.
2. Blueprints to achieving this will be discussed in Chapter 5.
3. The author's emphasis.

References

Bannister-Tyrrell, M; Harley, D & McMichael, T. (2015). *Detection and Attribution of Climate Change Effects on Infectious Diseases* (Australia: Australia National University Press, 2015) 447–459. 10.22459/HPPP.07.2015.25

Benett L, *Deforestation and Climate Change* (Climate Change Institute, Washington, DC, 2017).

Caminade C, McIntyre K and Jones A, 'Impact of Recent and Future Climate Change on Vector-Borne Diseases' [2019] 1436(1) *Annals of the New York Academy of Sciences* 157–173.

Carl-Johan N, 'How Urbanization Affects the Epidemiology of Emerging Infectious Diseases' [2015] 5(1) *Infection Ecology & Epidemiology* 27060.

Chan J et al, 'Interspecies Transmission and Emergence of Novel Viruses: Lessons from Bats and Birds' [2013] 21(10) *Trends in Microbiology* 544–555.

Chan N, *Urbanization, Climate Change and Cities: Challenges and Opportunities for Sustainable Development* (Keynote paper presented at the Asia-Pacific Chemical, Biological and Environmental Engineering Society (APCBEES) International Conference, University Sains Malaysia, Penang, 2017).

Daszak P, 'The Anatomy of Pandemics' [2012] 380 *Lancet* 1883–1884; 'Emerging Zoonoses: A One Health Challenge' [2020] 19 *Clinical Medicine* 100300.

Enuoh O and Bisong F, 'Biodiversity. Conservation & Commercial Bushmeat Handling Challenges in African Parks and Protected Areas: A Critical Review and Synthesis of Literature' [2014] 4(15) *Research on Humanities and Social Sciences* 39–57.

Estrada-Pena A et al, 'Effects of Environmental Change on Zoonotic Disease Risk: an Ecological Primer' [2014] 30(10) *Trends in Parasitology* 1269.

FAO, *Shaping the Future of Livestock: Sustainable, Responsibly, Efficiently* (10th Global Forum for Food and Agriculture (Rome 2018).

Friend M, McLean R and Dein F, 'Disease Emergence in Birds: Challenges for the Twenty–First Century' [2001] 118 *The Auk* 290–303.

Georgiou A, 'Risk of Virus Spill over from Animals to Humans Increases Through Wildlife Exploitation and Domestication' *Newsweek* (4 April 2020).

Greger M, 'The Human/Animal Interface: Emergence and Resurgence of Zoonotic Infectious Diseases' [2007] 33(4) *Critical Reviews in Microbiology* 243–299.

Hansungule M and Jegede A, 'The Impact of Climate Change on Indigenous Peoples' Land Tenure and Uses: The Case for a Regional Policy in Africa' [2014] 21 *International Journal on Minority and Group Rights* 256–291.

Helmholtz Centre for Environmental Research, *Deforestation in the Tropics* (Germany, 2018).

IPCC's Fifth Assessment Report, *Mitigation of Climate Change* (Washington, 2014).

Jay-Russel M and Doyle M, *Food Safety Risks from Wildlife: Challenges in Agriculture, Conservation and Public Health* (Springer, Switzerland, 2016).

Jones B, 'Zoonosis Emergence Linked to Agricultural Intensification and Environmental Change' [2013] 110(21) *PNAS* 8399–8404.

Kareh W et al, 'Ecology of Zoonoses: Natural and Unnatural Histories' [2012] 380 *Lancet* 1936–1945.

Losh J, 'A Battle to Protect Forests Unfolds in Central Africa' *New York Times* (28 January 2020).

Majumder M, *Impact of Urbanization on Water Shortage in Face of Climatic Aberration* (Springer, Geneva, 2015) 3–20.

Mantovani A, *Notes on the Development of the Concept of Zoonoses* (WHO Mediterranean Zoonoses Control Centre Information Circular, Greece, 2001) 51.

Padma T, 'How Natural Habitat Destruction Can Fuel Zoonotic Diseases Like COVID 19' *Mongabay India* (5 June 2020).

Pörtner H, Roberts DC, Masson-Delmotte V, Zhai P, Tignor M, Poloczanska E, Mintenbeck K, Nicolai M, Okem A, Petzold J, Rama B and Weyer N (eds), *IPCC Special Report on the Ocean and Cryosphere in a Changing Climate* (Cambridge: Cambridge University Press, 2019).

Robert ES and Oaks S Jr (eds), *Emerging Infections: Microbial Threats to Health in the United States* (National Academies Press (US), Washington, DC, 1992).

Roman-Palados C and Wiens J, 'Recent Responses to Climate Change Reveal the Drivers of Species Extinction and Survival' [2020] 117(8) *PNAS* 4211–4217.

Spec A, Escota G, Chrisler C and Davies B, *Comprehensive Review of Infectious Diseases* (Elsevier, New York, 2020).

Spickler A, 'Transmission of Zoonoses Between Animals and People' *Merck Veterinary Manual* (2015).

UNEP, 'Preventing the Next Pandemic – Zoonotic Diseases and How to Break the Chain of Transmission' (Nairobi, Kenya, 2015).

UNEP, CBD and WHO, *Connecting Global Priorities: Biodiversity and Human Health: A State of Knowledge Review* (Geneva, 2015).

United Nations Environment Programme, *Emerging Zoonotic Diseases and Links to Ecosystem Health – UNEP Frontiers, 2016 Chapter* (Nairobi: UNEP, 2020).

Vanwambeke S, Linard C and Gilbert M 2019 'Emerging Challenges of Infectious Diseases as a Feature of Land Systems' [2019] 38 *Current Opinion in Environmental Sustainability* 31–36.

Woo P, Lau S and Yuen K, 'Infectious Diseases Emerging from Chinese Wet-Markets: Zoonotic Origins of Severe Respiratory Viral Infections' [2006] 19(5) *Curr Opin Infect Dis* 401–407.

Woodward A, 'Both the New Coronavirus and SARS Outbreaks Likely Started in Chinese Wet Markets' *Business Insider* (January 23, 2020).

Woolhouse M and Gowtage–Sequeria S, 'Host Range and Emerging and Re-Emerging Pathogens' [2005] 11 *EID* 1842–1847.

World Health Organization, *The Constitution of the World Health Organization* (WHO: Geneva, 1946).

World Health Organization, *Constitution of the World Health Organization* (1946). Forty-Fifth Edition (Geneva: World Health Organization, 2006).

2 Why traditional ecological knowledge matters

Ecological stewardship is borne out of respect, justice, truth, sensitivity and compassion – the very pillars of traditional ecological knowledge.

Introduction

Starting in the 21st century, humanity has faced unprecedented dangers and problems such as air and water pollution, drought, disasters and most recently, increasing pandemics. As a result, most scientific work has gone into devising coping strategies to find solutions for these problems. Thus, with the failure of development theories in the last three decades, the focus in most of the social sciences has changed. Development theorists have begun using Indigenous Knowledge (IK) as an influential tool to enhance the process of sustainability (Meadows et al. 2019). Nevertheless, development theorists and scientists can be divided into two groups. One group believes that indigenous people with IK cannot change anything because they are trusting nature to make and determine provisions for all aspects of their survival and their future lives. As a result, they believe that IK cannot have any effective impact on sustainable development and natural resource management (Wallner 2005). Despite these views, a second group believes that IK contains legitimate knowledge and has the capacity to help the process of sustainable development and natural resource management (Makondo & Thomas 2018). However, current views about IK accept that IK is an important tool which holds promise for agriculture, food security and sustainable development and is able to provide alternative development approaches (Ogundiran 2019). The reason for changes in the views of theorists with regard to IK is based on two main points. Firstly, IK is capable of working with different trends in the social sciences in terms of the thinking and development of administrative practices and sustainability. Secondly, IK is reckoned to be an important natural resource that is able to facilitate the development process in terms of cost-effective and sustainable ways. This is because IK covers the whole range of human experience such as being able to integrate with physical sciences such as agriculture, medicine, climatology, engineering and irrigation or with social sciences such as politics, economics, military studies and sociology, or areas of humanities such as communications, arts and crafts, and so on (Denzin et al. 2008). IK is therefore knowledge that is unique to a given culture or society. It contrasts with the international knowledge system generated by universities, research institutions and private firms. It is the basis for local-level decision-making in agriculture, health care, food preparation, education, natural-resource management and a host of other activities in rural communities (Dobson 1999; Hopwood et al. 2005; Purvis 2019).

This chapter will discuss indigenous ecological knowledge, popularly known as traditional ecological knowledge, with a view to examining its nature and place in contributing to biodiversity conservation and/or restoration.

Traditional ecological knowledge of indigenous peoples

Fikret Berkes has developed a definition of *Traditional Ecological Knowledge* (TEK) as 'a cumulative body of knowledge, practice, and belief, evolving by adaptive processes and handed down through generations by cultural transmission, about the relationship of living beings (including humans) with one another and with their environment' (Berkes 2017). TEK is often used to describe the knowledge held by indigenous cultures specifically about their immediate environment and the cultural practices that build on that knowledge. It includes an intimate and detailed knowledge of plants, animals and natural phenomena, the development and use of appropriate technologies for hunting, fishing, trapping, agriculture and forestry, and a holistic knowledge, or 'world view' which parallels the scientific discipline of ecology (Berkes et al. 1998; Nelson & Shilling 2018).

Disciplines from anthropology and ethnobiology to systems ecology and resilience theory have demonstrated the contribution of TEK to improving livelihoods, sustaining biodiversity and ecosystems services, and building resilience in social-ecological systems (Denxzin et al. 2008). The potential contribution of TEK to building resilience in social-ecological systems has gained growing attention in the context of accelerated global change and generalized ecosystem service decline. Throughout history, communities maintaining tight links to ecosystem dynamics have developed knowledge, practices and institutions to accommodate recurrent disturbances to secure their livelihood. Because it co-evolves with ecological and social systems, TEK can strengthen the capacity of human societies to deal with disturbances and to maintain ecosystem services and under conditions of uncertainty and change (Gómez-Baggethun et al. 2013).

Since the advent of modernity, and most notably since the launch of the industrial revolution in Europe, expanded to other areas through the globalization process, TEK has been eroded in many parts of the world. Erosion of TEK systems owes to complex and multifaceted reasons, including the compounding influences of formal schooling and loss of local languages; dominant religions; changes in land use; market integration; loss of access to resources through conservation programs; mechanization of resource systems, and, more generally, industrialization and globalization processes (Rosenthal 2014). The increase of the scale and pace of global change since the so-called 'great acceleration' of the mid-20th century raised the question of whether TEK systems would adapt or disappear in the face of urbanization, technological development and market globalization. Over the second half of the 20th century, the decline in traditional lifestyles and associated knowledge was so widespread that when academia rediscovered TEK by the 1980s many doubted whether TEK systems would even survive the millennium (Berkes et al. 2000).

Over the last two decades, however, major developments are reshaping societal perceptions regarding the fate of TEK. First, in recent years, researchers are updating their perceptions of TEK's ability to adapt to change. Until recently, TEK was largely perceived as a vestige of the past that held – at best – folkloric interest and was bound to disappear with economic development. Yet, recent research from developed and developing countries has found that substantial pockets of TEK persist in many rural and urban areas that have been subject to modernization processes called *pockets of social-ecological memory*, those places that having captured, stored and transmitted through time the knowledge and experience of managing a local ecosystem and the services it produces, continue to maintain and foster them despite drastic changes in the surrounding environments. For example, agricultural landscapes in Europe have evolved through thousands of years of interactions between social and ecological systems and have drastically changed in the last century with societal transformation and industrialization of agriculture. Yet many places still preserve local and traditional farming knowledge and techniques.

Thus, the perception of TEK in the academia is shifting from one in which TEK was mainly perceived as existing in a rather essentialized and static form to one in which TEK is increasingly seen as having a hybrid and dynamic nature, more capable of adapting to new ecological and socioeconomic conditions than previously assumed (Berkes et al. 2000). The dynamic nature of TEK is sometimes achieved through the accommodation of new forms of knowledge and by disregarding those knowledge components that become obsolete or less useful for daily life, provided that local people maintain the capacity to apply their knowledge (Cordero & Krishnan 2018). For example, there are cases where local people maintain traditional practices on vegetable gardens but also adopt greenhouses, as those improve the conditions for certain crops. Within the same context, there may also be co-existence and complementarity of medicinal plants and pharmaceuticals knowledge among an indigenous population. Likewise, many indigenous societies have retained animistic elements and world views merged with the religions to which they have been converted, and these world views keep affecting their activities and relations with their environments (Inglis 1993; Paredes & Hopkins 2018).

Secondly, these new perspectives on the adaptive nature of TEK have favoured an increasing recognition of the value of such knowledge in environmental policy. International policy processes such as the United Nations' Declaration on the Rights of Indigenous Peoples (UNDRIP) and the Convention on Biological Diversity (CBD) have encouraged national governments to recognize and protect TEK for the conservation and sustainable use of biological diversity as well as to promote its wider application in resource management and biodiversity conservation (UNDRIP 2007; CBD 1992). This call has been taken up by national legislation in some countries that have started to develop national inventories of their TEK systems, acknowledging them as an important part of their cultural heritage. Other major international initiatives for the protection of ecosystem services and biodiversity, such as the Millennium Ecosystem Assessment and The Economics of Ecosystems of Biodiversity, have also stressed the importance of traditional societies and associated knowledge and value systems for biodiversity protection. Also, the Intergovernmental Platform on Biodiversity and Ecosystem Services (IPBES) has emphasized the importance of TEK in sustaining ecosystem services and biodiversity worldwide. A set of procedures for working with indigenous and local knowledge systems was approved at IPBES 4. It is thus expected that established and emerging policy recommendations will translate into the implementation of programs to sustain, protect and restore TEK, as well as its associated lifestyles.

Third, paralleling these trends, in some academic and civil society circles there is a mounting questioning of the techno-scientific rationality and economic growth ideology of industrial Western civilizations. These trends go hand in hand with a seemingly revalorization of traditional lifestyles and associated knowledge systems and world views. Scholars concerned with TEK and other components of biocultural diversity have unearthed features of small-scale societies that had been downplayed by the societal imaginary of modernity. These features include small-scale societies' capacity to harmonize livelihoods with biodiversity conservation; adoption of reciprocity motives to drive their economies; working time limited by needs and the capacity for collective action in governing common-pool resources (Gadgil et al. 1993). In essence, then, the status of TEK has been upgraded over the last two decades, not only in academia, but increasingly also among policymakers and civil society. Calls for the recognition of TEK are being slowly taken up by international treaties as an important potential contribution to the portfolio of responses to major social and environmental problems that humanity faces at present, including biodiversity loss, ecosystem service decline, and increased vulnerability and uncertainty associated with global environmental change (Ruiz-Mallén & Corbera 2013).

The respect and restorative ethic of traditional ecological knowledge

This book zeroes in on the *respect* and *restorative* ethic of indigenous peoples'[1] TEK and its significance for biodiversity conservation. TEK is borne out of rich traditions embedded in the ethics of protecting Nature. A review of various ancient cultures shows how communities lived in harmony with Nature, with a tradition of reverence for the elements that constitute ecosystems, drawing their sustenance from Nature; and at the same time protecting the ancestral lands from which they were born and nurtured for millennia. This is clearly manifested in many of the traditional practices, spiritual beliefs, rituals, customary laws, arts and crafts, from indigenous peoples (Berkes 1999; Brooks & Brooks 2010).

Respect for Nature is a universal practice in indigenous cultures across our planet. For indigenous peoples, life is a gift to be celebrated. They believe that Nature is not 'something' to be used, possessed or an object to be exploited, but a living entity and subject of reverence; and the relationship between humans and Nature should be that of sacred trust and love (Uprety et al. 2012). Indigenous peoples understand that consciousness and feelings are not only human attributes, but the whole of Nature is alive, animate, imbued with spirit or consciousness – plants, animals, rivers, lakes, mountains, the wind, sun and moon. Respect for Nature is inherent, rooted in their intimate relationship and understanding of Nature's living processes and participants (Taylor 1981). According to Thomas Berry, indigenous peoples derive their customary laws from Nature, recognizing that we humans need to comply with the laws of Nature for the well-being of all. His world view of indigenous peoples is reflected in Earth systems science, which recognizes the unity of all life on Earth as a living organism and obliges human beings to understand, participate and cooperate with our source of life (Berry 2000). An indigenous person's sense of self is not separate from the land. The interconnectedness with the land and the natural world is a lived experience. Indigenous persons have a hard time knowing themselves and being themselves without this relationship to their homeland. The vital knowledge of generations has taught them how to live with nature and be in balance and harmony with the natural world (Noisecat 2017).

Today, we see the consequences of violating the laws that govern life since the inception of the industrial period – climate chaos, extreme weather events such as flooding and landslides, ecosystem collapse, mass species extinction and growing hunger, poverty and ill health amongst humans. The industrial system has spread across the planet, with life-threatening consequences for all. This is triggering a growing recognition that indigenous *respect* and *restoration* ethos can help to show us the way forward – that it is possible for humans to live in harmony with Nature again if the right values are deployed (Kovach 2009).

The English language itself has played a significant role in holding us back. By imposing a classification system upon the world in which you are either a human or a thing, we isolate ourselves and distort our perspective to the extent that all else becomes an 'it' or a falsely engendered *he* or *she* (Kimmerer 2013). It is this disconnection that leads us down the path to destructive exploitation of the natural world. In our scientific approaches, we focus on breaking these *things* down to their smallest components so that we may better understand them, but in doing so we close ourselves off from the bigger picture, from the *subject* itself. Kimmerer and others see the need for a paradigm shift in scientific practice that changes from a study of *objects* to a study of *subjects* with agency and wisdom that is of great value to the world we share. Achieving such a shift in perception and practice is easier said than done, but a crucial step towards this goal may come in the form of vocabulary (Kimmerer 2013). Introducing the 'grammar of animacy' to our discussions of research *subjects* could open the door to new conceptual models for understanding the ways in which the inhabitants of this world are all interconnected. Chapter Three will demonstrate that an important way to move forward is to encourage open dialogue between mainstream science and TEK, but that dialogue cannot ignore the former abuses perpetuated in the name of mainstream science. Providing greater support for the continued development of indigenous research frameworks by (and for) indigenous scholars is crucial to the future of this relationship (Kimmerer 2013; Smith 2013).

These moral values that underpin TEK spring not simply from recognizing people's dependence on the natural world but rather from a deeper notion of unity with it. As one member of the Penobscot Nation has said, 'the river is us; the river is in our veins.... If the rivers are the lifeblood of our Mother and they become contaminated, then the same happens to us' (Long et al. 2020). Or as one Hopi Tribal leader has explained, 'This is the place that made us' (Casagrande & Vasquez 2009). Leslie Marmon Silko has noted that the term 'landscape' tends to reduce human beings to mere viewers when actually 'The land, the sky, and all that is within them – the landscape – includes human beings' (Silko 1987). Accordingly, many tribal perspectives may favour terms such as 'homeland' over 'landscape' when describing the aims of their restoration efforts.

World views of indigenous peoples are often rooted in 'kincentricity', with origin stories that depict humans as having been able to converse with 'four-legged' animals, 'winged animals' and other non-human creatures since their emergence (Aldren & Goode 2014). Tribes inform their present-day restoration efforts based upon inherited belief systems; for example, 'the Nium still consider these animals, plants, water, and wind to be relatives and full citizens in their community' (Aldren & Goode 2014). The world view that all living things are connected through living communities is expressed through norms that consider impacts on entire ecosystem networks. Traditional practices do often promote species that have high utilitarian value as food, medicine and materials that help people live, but they also consider how those food webs support non-human inhabitants (Aldren & Goode 2014).

Indigenous perspectives commonly regard water bodies, rocks and other places as alive and therefore worthy of thoughtful and respectful care (Bang et al. 2012). Indeed, research has found that native students regard this view as one of the most prominent differences between the teaching of elders and those of typical mainstream scientists (Bang et al. 2012). The concept that the natural world is imbued with life and our home, rather than merely a source of services, reinforces the idea that restoration is a way towards proper living rather than simply another tool to serve humans.

Academics have popularized the concepts of consilience and transdisciplinary science, which suggest that people can achieve deeper understandings by working across the boundaries and scale limitations of individual disciplines in natural sciences and humanities (Wilson 1998; Senos et al. 2006). TEK has also been suggested as fertile ground for encouraging boundary-spanning scientific endeavours (Berkes 2009; Pretty et al. 2009). Traditional perspectives suggest that restoration can be more conscientious by being careful, thoughtful and vigilant regarding right and wrong. Restoration can also be more consensual by upholding the principle that tribes and other indigenous communities should have 'free, prior and informed consent' (David-Chavez & Gavin 2018).

Conclusion

In spite of the modernization, traditional ecological ethos continues to survive in many other local societies, although often in reduced forms. Investigations into the traditional resource use norms and associated cultural institutions prevailing in rural societies have demonstrated that a large number of elements of local biodiversity, regardless of their use value, are protected by the local cultural practices. Some of these may not have known conservation effect, yet may symbolically reflect a collective appreciation of the intrinsic or existence value of life forms, and the love and respect for nature. Later (in Chapter Four) it will be demonstrated that traditional conservation ethics are still capable of protecting much of the world's decimating biodiversity, as long as the local communities have even a stake in the management of natural resources.

Note

1 *Indigenous Peoples* are ethnic groups who are native to a particular place on Earth and live or lived in an interconnected relationship with the natural environment there for many generations

References

Aldern JD and Goode RW, 'The Stories Hold Water: Learning and Burning in North Fork Mono Homelands' [2014] 3 *Decolonization: Indigeneity, Education and Society* 26–51.

Bang M, Warren B, Rosebery AS and Medin D, 'Desettling expectations in science education' [2012] 55 *Human Development* 302–318.

Berkes F, *Sacred Ecology: Traditional Ecological Knowledge and Resource Management* (Taylor & Francis, Philadelphia, 1999) 1–20.

Berkes F, 'Evolution of Co-Management: Role of Knowledge Generation Bridging Organizations and Social Learning' [2009] 90 *Journal of Environmental Management* 1692–1702.

Berkes F, *Sacred ecology* (Routledge, New York, 2017).

Berkes F, Colding J and Folke C, 'Rediscovery of Traditional Ecological Knowledge As Adaptive Management' [2000] 10 *Ecological Applications* 1251–1262.

Berkes F, Kislalioglu M, Folke C and Gadgil M, 'Exploring the Basic Ecological Unit: Ecosystem-Like Concepts in Traditional Societies' [1998] 1 *Ecosystems* 409–415.

Berry T, *The Great Work: Our Way into the Future* (Broadway Books, New York, 2000).

Brooks L and Brooks C, 'The Reciprocity Principle and Traditional Ecological Knowledge' [2010] 3(2) *International Journal of Critical Indigenous Studies* 11–28.

Casagrande DG and Vasquez M, 'Restoring for Cultural–Ecological Sustainability in Arizona and Connecticut' in M Hall (ed), *Restoration and History: The Search for a Usable Environmental Past* (Routledge Press, New York, 2009) 195–209.

Converntion on Biological Diversity (CBD) (Montreal: United Nations, 1992).

Cordero R and Krishnan M, 'Elements of Indigenous Socio-Ecological Knowledge Show Resilience Despite Ecosystem Changes in the Forest-Grassland Mosaics of the Nilgiri Hills, India' [2018] 4 *Palgrave Communications* 105.

David-Chavez DM and Gavin MC, 'A Global Assessment of Indigenous Community Engagement in Climate Research' [2018] 13 (12) *Environmental Research Letters* 13.

Denzin N, Lincoln Y and Smith L, *Handbook of Critical and Indigenous Methodologies* (Sage Publishing, CA, 2008).

Gadgil M, Berkes F and Folke C, 'Indigenous Knowledge for Biodiversity Conservation' [1993] 22 (2–3) *Ambio* 151–156.

Gómez-Baggethun E, Corbera E and Reyes-García V, 'Traditional Ecological Knowledge and Global Environmental Change: Research Findings and Policy Implications' [2013] 18(4) *Ecology and Society* 72.

Hopwood B, Mellor M and O'Brien G, 'Sustainable Development: Mapping Different Approaches' [2005] 13(1) *Sustainable Development* 38–52

Inglis J, *Traditional Ecological Knowledge: Concepts and Cases* (Canadian Museum of Nature, Ottawa, 1993).

Jacobs M, 'Sustainable Development as a Contested Concept' in A Dobson (ed), *Fairness and Futurity* (Oxford University Press, Oxford, 1999) 21–45.

Kimmerer Robin, *Braiding sweetgrass: Indigenous wisdom, scientific knowledge and the teachings of plants* (Milkweed Editions, Minessota, 2013) 56–57.

Kovach M, *Indigenous methodologies: Characteristics, conversations, and contexts* (University of Toronto Press, Toronto, 2009).

Long J et al, 'How Traditional Tribal Perspectives Influence Ecosystem Restoration' [2020] 12(2) *Ecopsychology* 71–82.

Makondo C and Thomas D, 'Climate Change Adaptation: Linking Indigenous Knowledge with Mainstream Science for Effective Adaptation' [2018] 88 *Environmental Science and Policy* 83–91.

Meadows J, Annadale M and Ota L, 'Indigenous Peoples Participation in Sustainability Standards for Extractives' [2019] 88 *Land Use Policy* 104118; Antwi J, 'The Indigenous Ecological Knowledge and the Environment: the Akan Perspective' [2020] 2 (1) *MOTBIT* 58–69.

Nelson M and Shilling D, *Traditional Ecological Knowledge: Learning from Indigenous Practices for Environmental Sustainability* (Cambridge University Press, New York, 2018).

Noisecat J, 'The Western Idea of Private Property is Flawed' *The Guardian* (27 March 2017) <https://www.theguardian.com/ commentisfree/ 2017/mar/27/wes tern-idea-private-property-flawed-indigenous-peoples-have-it-right> accessed 10 June 2020.

Ogundiran A, 'Food Security, Food Sovereignty and Indigenous Knowledge' [2019] 36 *African Archaeological Review* 343–346.

Paredes R and Hopkins A, 'Dynamism in Traditional Ecological Knowledge: Persistence and Change in the Use of *Totora* (*Schoenoplectus californicus*) for Subsistence in Huanchaco, Peru' [2018] 9(2) *Ethnobiology Letters* 169–179.

Pretty J, Adams B, Berkes F, De Athayde SF, Dudley N, Hunn E and Pilgrim S, 'The Intersections of Biological Diversity and Cultural Diversity: Towards Integration' [2009] 7 *Conservation and Society* 100–112.

Purvis B, Mao Y and Robinson D, 'Three pillars of sustainability: in search of conceptual origins' [2019] 14 *Sustainability Science* 681–695.

Rosenthal M, 'When Languages Die, Ecosystems Often Die with them' *Living on Earth* (2014).

Ruiz-Mallén I and Corbera E, 'Community-based conservation and traditional ecological knowledge: implications for social-ecological resilience' [2013] 18(4) *Ecology and Society* 12.

Senos R, Lake FK, Turner N and Martinez D, 'Traditional Ecological Knowledge and Restoration Practice in the Pacific Northwest' in D Apostol and M Sinclair (eds), *Restoring the Pacific Northwest: The Art and Science of Ecological Restoration in Cascadia* (Island Press, Washington, DC, 2006) 393–426.

Silko LM, 'Landscape, History, and the Pueblo Imagination' in D Halpern (ed), *On Nature: Nature, Landscape, and Natural History* (North Point, San Francisco, CA, 1987) 83–94.

Smith L, *Decolonizing methodologies: Research and indigenous peoples* (Zed Books Ltd, London, 2013).

Taylor P, 'The Ethics of Respect for Nature' [1981] 3 *Environmental Ethics* 197–218; Light A, 'Restorative Relationships' in R France (ed), *Healing Nature, Repairing Relationships: Landscape Architecture and the Restoration of Ecological Spaces* (MIT Press, Cambridge, MA, 2006).

United Nations Declaration on Rights of Indigenous Peoples (UNDRIP), 2007.

Uprety Y, Asselin H, Bergeron Y, Doyon F and Boucher J, 'Contribution of Traditional Knowledge to Ecological Restoration: Practices and Applications' [2012] 19(3) *Ecoscience* 225–237.

Wallner F, 'Indigenous Knowledge and Mainstream Science: Contradictions or cooperation' [2005] 4(1) *African Journal of Indigenous Knowledge Systems* 46–53.

Wilson EO, 'Consilience among the Great Branches of Learning' [1998] *Daedalus* 127(1) 131–149.

3 Complementarity of traditional ecological knowledge and mainstream science

The path to effective sustainable risk management is in learning by doing.

Introduction

When land is owned, managed or occupied in a traditional way, the word 'traditional' refers to a knowledge that stems from centuries-old observation and interaction with nature. In the preamble to this book, it was shown that this knowledge is often embedded in a cosmology that reveres the *one-ness* of life, considers nature as sacred and acknowledges humanity as a part of it. And it encompasses practical ways to ensure the balance of the environment in which they live, so it may continue to provide services such as water, fertile soil, food, shelter and medicines. This chapter reveals that the possibility of learning from the values, epistemologies and practices of non-Western cultures has not been lost on scholars dealing with complex organic systems and adaptive management – a systematic approach for improving resource management by learning from management outcomes.

Mainstream science and management have a questionable record with regard to long-term sustainability, whereas some indigenous or traditional peoples have developed systems that seem more sustainable than our own. Traditional resource management systems may thus be viewed as experiments in successful living and drawing upon knowledge of these alternatives may provide insights and speed up the process of adaptive management.

Mainstream science versus TEK

As ways of knowing, Western and indigenous knowledge share several important and fundamental attributes. Both are constantly verified through repetition and verification, inference and prediction, and empirical observations and recognition of pattern events.

While some actions leave no physical evidence (for example, clam cultivation) and some experiments can't be replicated (for example, cold fusion), in the case of indigenous knowledge, the absence of 'empirical evidence' can be damning in terms of wider acceptance.

Some types of indigenous knowledge simply fall outside the realm of prior Western understanding. In contrast to Western knowledge, which tends to be text based, reductionist, hierarchical and dependent on categorization (putting things into categories), indigenous science does not strive for a universal set of explanations but is particularistic in orientation and often contextual (Mistry & Berardi 2016).

One key attribute of mainstream science is developing and then testing hypotheses to ensure rigor and replicability in interpreting empirical observations or making predictions. Although hypothesis testing is not a feature of TEK, rigor and replicability are not absent (Paige & Stricker 2010).

Whether or not traditional knowledge systems and scientific reasoning are mutually supportive, even contradictory lines of evidence have value. Employing TK-based observations and explanations within multiple working hypotheses ensures consideration of a variety of predictive, interpretive or explanatory possibilities not constrained by Western expectation or logic (Huntington 2000). And hypotheses incorporating traditional knowledge-based information can lead the way towards unanticipated insights. Indigenous peoples don't need mainstream science to validate or legitimate their knowledge system. Some do appreciate the verification, and there are partnerships developing worldwide with indigenous knowledge holders and Western scientists working together (Huntington 2000).

In relation to scientists increasingly appreciating the inherent limitations of the classical scientific way of analyzing nature, or ecological systems, reference was made earlier to the traditional ecological knowledge-based approach that denies the usefulness of a reductionist approach to seeking cause and effect as an operational principle for serious enquiry. In this regard, it is as well to consider that scientists now also understand that, at a fundamental level, certain phenomena are best understood not as being composed of isolated entities that can be studied as such, but rather they can better be understood by means of the influence they have on other phenomena. In other words, by means of their systemic relationships, outside of which they in fact cease to be definable (Garnge 2004).

So fundamental is this realization to some leading thinkers, that one such, Gregory Bateson, has argued that relationships should be used as a basis for all definitions, and this should be taught to our children in elementary school. Anything should be defined not by what it is in itself, but by its relations to other things (Bateson 2002). It appears then, that in some important respects, the leading edge of scientific thinking is coming into remarkable alignment with the TEK-based system of understanding what is the appropriate way of comprehending nature (Kimmerer 2002).

Traditional ecological knowledge plays a significant role in monitoring, responding to and managing ecosystem processes and functions, with special attention to ecological resilience. TEK comprises local practices for ecosystem management; these include multiple species management, resource rotation, succession management, landscape patchiness management and other ways of responding to and managing pulses and ecological surprises (Berkes et al. 2000). Social mechanisms behind these traditional practices include a number of adaptations for the generation, accumulation and transmission of knowledge; the use of local institutions to provide leaders/stewards and rules for social regulation; mechanisms for cultural internalization of traditional practices; and the development of appropriate world views and cultural values (Ibid). Some traditional knowledge and management systems were characterized by the use of local ecological knowledge to interpret and respond to feedback from the environment to guide the direction of resource management. These traditional systems had certain similarities to adaptive management with its emphasis on feedback learning, and its treatment of uncertainty and unpredictability intrinsic to all ecosystems (Ibid).

A trajectory of adaptive management

Adaptive management, an approach for simultaneously managing and learning about natural resources, has been around for several decades. Interest in adaptive decision-making has grown steadily over that time, and by now many in natural resources conservation claim that adaptive management is the approach they use in meeting their resource management responsibilities (Williams 2011).

Adaptive management is based on the major premise that knowledge of ecological systems is not only incomplete but elusive (Walters & Holling 1990). Moreover, there is a growing conviction that expanding knowledge through traditional scientific enquiry will always be limited by resources and time. When these limiting factors are linked to the contextual conditions of resource scarcity, potential irreversibility and growing demands, the need for new ways in which understanding and learning not only occur but directly inform decision-making and policy processes becomes apparent. Adaptive management offers both a scientifically sound course that does not make action dependent on extensive studies and a strategy of implementation designed to enhance systematic evaluation of actions (Moore & McCarthy 2010; Plummer et al. 2012).

There are many definitions of adaptive management. The widespread use of the term has propagated various interpretations of its meaning and, as a result, there are only vague notions about what it is, what is required for it to be successful or how it might be applied. Not surprisingly, given recent attention by the scientific community, many definitions frame the discussion around a structured process that facilitates learning by doing, that is, 'adaptive management does not postpone action until "enough" is known, but acknowledges that time and resources are too short to defer *some* action' (Holling 1978; Walters 1986; Lee 1999).

The concept of learning is central to adaptive management and is grounded in recognition that learning derives from action and, in turn, informs subsequent action. Lee (1999) argues that the goal of implementing management experiments in an adaptive context is to learn something; he also argued that surprise is an inevitable consequence of experimentation and that it is often a source of insight and learning. Yet, such observations beg the question as to what learning is.

Michael (1995) argues that there are two kinds of learning: one for a stable world and one for a world of uncertainty and change. In a world of rapid change and high uncertainty, acquiring more facts – data – might not be as important as improving the capacity to learn how to learn. In other words, what might have once facilitated learning might no longer do so.

Adaptive environmental management

Adaptive Environmental Management (AEM) refers to the emerging directions which can be seen to be developing through the integration of ecological and participatory research approaches.

Many contemporary research efforts are concentrating on creating new approaches to more closely link science, management and policy at an ecosystem level. At the base, these efforts represent a search for a research and development model and practice which combine the features of:

- management-based experimentation and innovation;
- natural resource system management on scales larger than individual enterprises and communities;
- methods for bringing about capacity for action among multiple agencies and actors (with typically divergent, not to say antagonistic points of view and interests);
- facilitation of the social processes and organizational capacity to accomplish these.

AEM focuses on learning and adapting through partnerships of managers, scientists and other stakeholders who learn together how to create and maintain sustainable ecosystems. It helps managers maintain flexibility in their decisions, knowing that uncertainties exist and so provides the latitude to adjust direction to improve progress towards desired outcomes (Fernández-Giménez et al. 2019; Berkes 2009).

Effective implementation of AEM must therefore involve the active involvement and support of the full set of partners and stakeholders – that is, it must be as collaborative as possible. An inclusive approach is required not only to build understanding, support, credibility and trust among constituent groups, but also to ensure adequate problem-framing and access to the knowledge, experience and skills held by these groups. Because environmental conservation problems are social in origin and potential solutions are framed in a social context, effective management programs must therefore be designed in biophysical and social contexts. These contexts inform stakeholder management priorities and help stakeholders evaluate trade-offs in management decision-making. For example, a local (traditional) management context helps explain why stakeholders would prioritize learning about drought resilience by keeping cattle in a single large herd, rather than splitting them into several smaller herds. This will allow for the development of a drought forage reserve by resting several pastures, even though it will be associated with a reduction in daily livestock gains. This temporal environmental context would influence how stakeholders value learning about drought resilience over potential short-term financial gain from reduced stock density.

Therefore, acknowledging that AEM is shaped by social and environmental contexts at multiple scales may provide a pathway for the application of AEM-produced knowledge. The use of new knowledge depends on its salience, credibility and legitimacy to users (Beier et al. 2017). In multistakeholder collaborations, prolonged engagement and mutual knowledge exchange is often required to develop respect, trust, and ultimately the credibility and legitimacy of co-produced knowledge. A shared culture of learning will contribute to co-produced knowledge driven by stakeholder questions relevant to the local natural resource management context (Fabricius & Cundill 2014).

Linking TEK and adaptive environmental management: a new approach to science

There are several similarities between traditional or indigenous management systems and adaptive management. If the orderly and rational science of the Age of Enlightenment is replaced by a new paradigm along the lines of adaptive management, the chasm between indigenous knowledge and mainstream science essentially evaporates (Cajete 2000).

Many of the prescriptions of traditional knowledge and practice are consistent with AEM as an integrated method for resource and ecosystem management (Berkes et al. 2000). AEM, like many TEK systems, emphasizes processes that are part of ecological cycles of renewability and regards human use of the environment in terms of how well it fits these cycles. Like many TEK systems, AEM considers change as inevitable and assumes that nature cannot be controlled and yields cannot be predicted. In both AEM and TEK, uncertainty and unpredictability are considered to be characteristics of all ecosystems, including managed ones; in both, social learning appears to be the way in which societies respond to uncertainty. Often this involves social learning at the level of society or institutions (Ibid). The existence of (mainstream science) practices such as monitoring resource abundance, multiple species management and watershed-based management practices in some TEK systems is further evidence of the similarity between TEK and adaptive management. This makes a collaborative platform between mainstream science and TEK imperative.

A good example of this fluidity between mainstream science and TEK is rainwater harvesting, which has been found to be scientific and useful for rainfed areas. For instance, a validation comes from the Negev. Ancient stone mounds and water conduits are found on hillslopes over large areas of the Negev desert. Field and laboratory studies suggest that ancient farmers were very efficient in harvesting water. A comparison of the volume of stones in the mounds to the volume of surface stones from the surrounding areas indicates that the ancient farmers removed only stones that had rested on the soil surface and left the embedded stones untouched. According to results of simulated rainfall experiments, this selective removal increased the volume of runoff generated over one square meter by almost 250% for small rainfall events compared to natural untreated soil surfaces (Ghimire & Johnston 2015).

There is clearly much to be learned not only from a study of other cultural practices, but from other cultural ways of knowing as well (Fujitani et al. 2017). As Bateson once suggested with regard to contemporary ecological problems, it is possible that some of the most disparate epistemologies which human culture has generated may give us clues as to how we should proceed. Further, other attitudes and premises – other systems of human 'values' – have governed man's relation to his environment and his fellow man in other civilizations and at other times. In other words, our way is not the only possible human way. It is conceivably changeable (Bateson 2002).

AEM does not postpone actions until 'enough' is known about a managed ecosystem, but rather is designed to support action in the face of the limitations of scientific knowledge and the complexities and stochastic behaviour of large ecosystems. It aims to enhance scientific knowledge and thereby reduce uncertainties. Such uncertainties may stem from natural variability and stochastic behaviour of ecosystems and the interpretation of incomplete data, as well as social and economic changes and events (e.g., demographic shifts, changes in prices and consumer demands) that affect natural resources systems. AEM aims to create policies that can facilitate the response to deleterious ecological situations and challenges incidental to them, such as global pandemics. Instead of seeking precise predictions of future conditions, AEM recognizes the uncertainties associated with forecasting future outcomes, and calls for consideration of a range of possible future outcomes. Thus, policies formulated from AEM methods are designed to be flexible and are subject to adjustment in an iterative, social learning process (National Research Council 2004).

Conclusion

Biodiversity conservation cannot afford to be the subject of just any single body knowledge such as mainstream science, but it has to take into consideration the plurality of knowledge systems. There is a more fundamental reason for the integration of knowledge systems. Application of scientific research and local knowledge contributes both to the equity, opportunity, security and empowerment of local communities, as well as to the sustainability of the natural resources in general, and biodiversity in particular. Local knowledge helps in scenario analysis, data collection, management planning, designing of the adaptive strategies to learn and get feedback, and institutional support to put policies into practice. Science, on the other hand, provides new technologies, or helps in improvement to the existing ones. It also provides tools for networking, storing, visualizing and analyzing information, as well as projecting long-term trends so that efficient solutions to complex problems can be obtained.

Local knowledge systems have been found to contribute to sustainability in diverse fields such as biodiversity conservation and maintenance of ecosystems services, tropical ecological and biocultural restoration, sustainable water management, genetic resource conservation and management of other natural resources. Local knowledge has also been found useful for ecosystem restoration and often has ingredients of adaptive environmental management.

AEM has emerged as an avenue for natural resource management that accepts there are gaps in the knowledge of these systems, but enacts a constant feedback mechanism. This learning and decision-making cycle allows for the decision-maker to continue to learn more about the systems for subsequent ecological situations. Generally, adaptive management is also described as 'learning by doing'. This may be conflicting in many contexts as it basically opposes the precautionary principle. Here is where indigenous knowledge comes into play – local people have been learning by doing for centuries.

Without scientific research, low resources and no computers to gather and store data, the only way of acquiring and transmitting knowledge was by literally doing something, seeing how it goes and communicating its success or failure. This trial and error system was repeated innumerable times by ancient communities, who slowly became masters of their natural surroundings. So what is proposed for the future of adaptive management is that before we 'learn by doing' we should 'learn from the ones who learnt by doing'.

There are no doubts indigenous groups that burnt huge forests and killed many species before learning how to manage their environment. But the knowledge was eventually acquired and perfected for many years and is still available for us to use. Nowadays, we cannot afford trial and error practices in the large scale, so we must learn from the ones who did it back in the day before it is too late and the knowledge dies with the people.

References

Bateson G, *Mind and Nature: A Necessary Unity* (Hampton Press, New Jersey, 2002).

Berkes F, 'Evolution of Co-Management: Role of Knowledge Generation, Bridging Organizations and Social Learning' [2009] 90 *Journal of Environmental Management* 1692–1702.

Berkes F, Colding J and Folke C, 'Rediscovery of Traditional Ecological Knowledge as Adaptive Management' [2000] 10(5) *Ecological Applications* 1251–1262.

Byers BA, Cunliffe RN and Hudak AT, 'Linking the Conservation of Culture and Nature: A Case Study of Sacred Forests in Zimbabwe' [2001] 29 *Human Ecology* 187–218.

Cajete G, *Native Science: Natural Laws of Interdependence* (Clear Light Publishers, Santa Fe, NM, 2000).

Deb D and Malhotra KC, 'Conservation Ethos in Local Traditions: The West Bengal Heritage' [2001] 14 *Society and Natural Resources* 711–724.

Fabricius C and Cundill G, 'Learning in Adaptive Management: Insights from Published Practice'[2014] 19 (1) *Ecology and Society* 29.

Fernández-Giménez M, Augustine L, Porensky H, Wilmer J, Derner D, Briske M and Stewart O, 'Complexity Fosters Learning in Collaborative Adaptive Management' [2019] 24(2) *Ecology and Society* 29.

Fujitani M, McFall A, Randler C and Arlinghaus R, 'Participatory Adaptive Management Leads to Environmental Learning Outcomes Extending Beyond the Sphere of Science' [2017] 3 *Science Advances*, 1–11.

Garnge Le L 2004, 'Mainstream Science and Indigenous Knowledge: Competing Perspectives or Complimentary Frameworks' [2004] 18(3) *South African Journal of Higher Education* 82–91.

Ghimire S and Johnston J, *Traditional Knowledge of Rainwater Harvesting Compared to Five Modern Case Studies* (Environmental and Water Resources Congress, 2015)

Holling CS, *Adaptive Environmental Assessment and Management* (John Wiley, London, 1978) 377.

Huntington H, 'Using Traditional Ecological Knowledge in Science: Methods and Applications' [2000] 10 *Ecological Applications* 1270–1274.

Kimmerer R, 'Weaving Traditional Ecological Knowledge into Biological Education: A Call to Action' [2002] 52 *BioScience* 432–438.

Lee KN, 'Appraising Adaptive Management' [1999] 3(2) *Conservation Ecology* 3.

Michael DN, 'Barriers and Bridges to Learning in a Turbulent Human Ecology' in LH Gunderson, CS Holling and SS Light (eds), *Barriers and Bridges to the Renewal of Ecosystems and Institutions* (Columbia University Press, New York, 1995) 461–488; Armitage D, Marschke M and Plummer R, 'Adaptive Co-Management and the Paradox of Learning' [2008] 18 *Global Environmental Change* 86–98.

Mistry J and Berardi A, 'Bridging Indigenous and Scientific Knowledge: Local Ecological Must be Placed at the Centre of Governance' [2016] 352(6291) *Science* 1274–1275.

Moore AL and McCarthy MA, 'On Valuing Information in Adaptive-Management Models' [2010] 24(4) *Conservation Biology* 984–993.

National Research Council, *Adaptive Management for Water Resources Project Planning* (National Academies Press, Washington, DC, 2004) 19–21.

Plummer R, Crona B, Armitage DR, Olsson P, Tengö M and Yudina O, 'Adaptive Comanagement: A Systematic Review and Analysis' [2012] 17(3) *Ecology and Society* 11

Pyone K, 'Indigenous Knowledge in Natural Resource Management: Integrating Local Perspective into Conservation Strategies' [2019] 83 *Yale Environment Review*, 297–307.

Schmidt Paige P and Stricker H, *What Tradition Teaches: Indigenous Knowledge Complements Western Wildlife Science* (USDA National Wildlife Research Center - Staff Publications, 2010) 1283.

Walters CJ, *Adaptive Management of Renewable Resources* (Macmillan, New York, 1986) 374.

Walters CJ and Holling CS, 'Large-Scale Management Experiments and Learning by Doing' [1990] 71(6) *Ecology* 2060–2068 in G Stankey, Clark R and Bormann B, *Adaptive Management of Natural Resources: Theory Concepts and Management Institutions* (United States Department of Agriculture, Washington, DC, 2005).

Williams B, 'Adaptive Management of Natural Resources – Framework and Issue' [2011] 92(5) *Journal of Environmental Management* 1346–1353.

4 Restoring land, community and health

> Restoration of Nature requires an abandonment of a silo ideology for a meeting of ancient and modern minds in order to achieve conservation ideals.

Introduction

In Chapter One, it was emphasized that the planet is in immense ecological stress and human societies currently face unprecedented challenges such as climate change, species extinctions, pollution and deforestation, to name a few. Natural ecosystems are being altered beyond their capacity and the dynamic ecological patterns and processes that have been created over the last 4.6 billion years are now at risk. Green infrastructure, ecological conservation, carbon sequestration, carbon emission reduction projects, ecological restoration and other technologies have been implemented to mitigate the global ecological crisis we face. Chapter Two showed that there is however a fundamental, missing piece that is severely underutilized and considered in our dire state. It involves a very basic concept of combining traditional ecological knowledge and indigenous world views with modern mainstream science (MS) to create environmentally sustainable solutions. The wealth of knowledge about the local environment within tribal communities is vast and diverse. It has developed over thousands of years and been passed down through a multitude of generations in oral teachings. Chapter Three has shown that TEK complements mainstream science and is increasingly being recognized by natural resource managers and scientists throughout the world as adaptive environmental management. On the other hand, mainstream science can be credited with numerous innovations and technological advances in fields ranging from engineering to medicine to natural resource management. This chapter seeks to show that combined, these two knowledge sources may provide powerful solutions to our most dynamic and complex environmental problems, especially the loss of biodiversity that is at the root of (present and future) pandemics.

The growing application of traditional ecological knowledge

As explained in Chapter Two, Traditional Ecological Knowledge (TEK) includes indigenous and local ecological knowledge that have been used to refer to sources of knowledge about species, ecosystems or practices held by people whose lives are closely linked to their natural environment.

TEK can not only add to an existing body of scientific knowledge, but can present a completely different picture of reality, especially when held within a different cosmological and ethical framework. When knowledge about the consequences of management is scarce, these alternative narratives can be of great value. The demonstrated complementarity between traditional and scientific sources of information has validated the use of TEK in ecological research.

With the planet losing species 100–1,000 times faster than the natural extinction rate, international experts assembling for high-level global biodiversity meetings, such as the recent UN Biodiversity Summit held in September 2020 and the Intergovernmental Science-Policy Platform on Biodiversity and Ecosystem Services (IPBES), agree that indigenous peoples should be engaged in arresting biodiversity loss more than ever. The United Nations released two major reports recognizing the leadership of indigenous peoples in sustaining our planet. The reports look at the fragile state of Nature. Together they illustrate what place-based peoples have known all along: everything is interrelated. If we burn through natural resources, we undermine our own futures. Yet if we relate to the natural world in a thoughtful way, the beings from the land and water such as herring, salmon, caribou and moose will remain – not just for us but for generations to come. They call for scaling up solutions, including expanding protected areas and strengthening the role of indigenous peoples, in biodiversity strategies. The reports find that one of the main reasons the world has failed to meet its biodiversity targets is the failure to recognize the vital contributions indigenous peoples and local communities make to conservation (CBD 2020).

IPBES experts have also agreed that knowledge co-production with indigenous peoples has growing importance with respect to managing biodiversity loss (International Institute for Sustainable Development 2013). Indeed, they note that processes that merge multiple sources and types of knowledge already help manage challenges as diverse as wildfires and animal herds (International Institute for Sustainable Development 2013).

Communities that are dependent on natural resources can rapidly develop insight into factors influencing resource availability or quality. Such information can be shared among users and can develop into a substantial body of knowledge (Rist et al. 2010). The use of TEK has found favour in conservation planning and resource assessment for three reasons: efficiency, additionality and community engagement (Pierotti & Wildcat 2000; Sheil & Lawrence 2004; Drew 2005). When TEK corresponds well to scientific data, it can be a more efficient method of acquiring information. Although such bodies of knowledge develop over significant periods of time and represent considerable investment by knowledge holders in experimentation and observation, rigorous social science methods can often gather some of this information in less time and at less cost than formal ecological research (Rist et al. 2010). Resource users often interact with a landscape at a much larger scale and over longer periods of time than are possible in standard scientific investigations (Wehi 2009). In addition, programs that garner the support of local people through their participation have a greater chance of acceptability and therefore long-term sustainability (Schwartzman et al. 2000; Bowen-Jones & Entwistle 2002; Danielsen et al. 2005).

The following sections discuss the growing complementarity or potential thereof between mainstream science and TEK. It will give examples from most regions of the world where studies have used TEK effectively to address conservation aims, sustainable resource use and climate change.

Africa

TEK in small-scale agricultural systems in Malawi

Farmers in Malawi classify soils according to their fertility levels in order to deploy suitable and appropriate agricultural management practices. In assessing soil quality, all the farmers in the study classified them, firstly, in relation to the length of time of continuous cultivation of a given field, and, secondly, in relation to soil properties, primarily the soil's 'slippery' or coarse texture (Tchale et al. 2005). The farmers' determination of soil texture resembles scientific ways of classification, which interestingly is also based on the feel method. Rubbing soil between the fingers to determine soil texture is both a scientific as well as an indigenous way of determining its properties. Soils in virgin gardens (*mphangula*), irrespective of their actual properties, are considered to be rich in soil nutrients before cultivation is undertaken.

Scientists have established that *Brachystegia* is a nitrogen-fixing plant. Farmers are generally not aware of this particular scientific finding, however, and certainly may not understand the process of nitrogen fixing in the soil by the *Brachystegia*, but they are well aware of the high soil fertility levels in those areas where the species is dominant. The species' ability to fix nitrogen in the soil therefore explains the high fertility levels that farmers attach to it through their observation of the growth of crops on those soils that have had such trees on them before conversion to farmland (Roth 2001).

Clearly, local knowledge complements Western scientific ways of knowing. The presence of *Brachystegia* removes the need to take soil samples just for the sake of determining soil fertility levels in choosing farm sites, which is an expensive and time-consuming process that requires the services of technicians.

Indigenous weather forecasting for climate resilience

Farmers and pastoralists in East Africa have relied on the indigenous weather prediction methods for generations. The great majority of farmers and pastoralists, particularly in Ethiopia's Afar and Borana sites, depend heavily on indigenous weather information compared to Tanzania and Uganda. Different weather forecasting indicators are used across the region.

Meteorological and astrological indigenous weather forecasting indicators include direction and strength of winds, star-moon alignment, apparent movement of stars, direction of the moon crescent, types of clouds, temperature conditions, lightning and thunder, colour of the sky, and rainbow to forecast the next rainy season. In Uganda, indigenous knowledge (IK) forecasters associate the onset of rainfall with the appearance of clouds. The appearance of nimbostratus and cumulonimbus clouds indicates a high probability of rainfall. While in Ethiopia, the appearance of a white feather-like column (vertically standing) cloud in the sky is an indication that the rain is about come. A sky dominantly covered with light clouds indicates drought (Mahoo et al. 2015).

Biological indicators focus on the behaviour and activities of domestic and wild animals, insects and different species of plants for weather forecasting. For instance, in Uganda the *Mvule* tree indicates onset of the rainy season. In Ethiopia, the intestines of cattle, sheep and goats are used to forecast the magnitude, severity, and duration of drought, drought-affected places, disease outbreak, the prospect of peace and/or conflict. In Tanzania, the occurrence of large flocks of swallows and swans, roaming from the South to the North during the months of September to November, is an indication of onset of short rains (Mahoo et al. 2015).

Different indigenous weather forecasting indicators have varying levels of acceptance depending on their precision. For instance, star-moon alignment is the most dependable weather information sources in Borana, while cloud colour is the preferred indicator in Hoima and Rakai sites. The challenges facing indigenous knowledge weather forecasting include insufficient documentation of the knowledge and a poor knowledge transfer system, lack of coordinated research to investigate its accuracy and reliability, death of forecast experts, and influence of religion and modern education.

These indigenous weather forecasting methods will no doubt play a major role in local livelihoods and are crucial to supporting local efforts to forecast and make sense of seasonal climate situations at the local level. However, progressive loss of indigenous knowledge threatens the ability of pastoralists and farmers to cope with and adapt to climate change. It is therefore imperative to find ways of integrating indigenous weather forecasting with the scientific weather forecasting systems.

Asia

Sustainable management of marine resources in the Pacific Islands

Sustainable management of marine resources, as practiced by many Pacific Island communities, traditionally involves the use of area and time-based restrictions to facilitate marine resource recovery. These traditional management systems involve a range of strategies, including tabu areas (sacred sites), species-specific prohibitions, seasonal and area closures to create networks of refuges, gear restrictions, behavioural prohibitions, totemic restrictions and food avoidance – all promoting a balanced approach to resource management (Jupiter et al. 2014).

The South Pacific is home to an array of marine life, including humpback whales, large schools of tuna, saltwater crocodiles, sharks and globally threatened turtles. This region is also home to a vast diversity of people, cultures, languages and traditions. But the marine resources of this island region are threatened by major challenges, including:

- destructive fishing;
- sea level rise;
- growing populations;
- coastal development;
- general loss of cultural/traditional connections to the sea.

Local people in the South Pacific have experience using traditional management systems for marine resources, such as seasonal bans and temporary no-take areas. These techniques draw on spiritual beliefs and resource management practices. Some coastal communities also combine traditional methods with modern techniques to implement the most effective forms of sustainable management. The fact that communities often retain either legal or de facto community tenure over local marine resources provides incentives for conservation efforts and ensures that communities receive the resulting benefits.

Restoring land, community and health 41

These community-based marine protection practices vary across countries and cultures throughout the area, but collectively have become known as the Locally Managed Marine Areas (LMMA) approach.

The resource management system of Pacific Islanders has strong potential to formulate better and more applicable resource management arrangements. With the continued failure of contemporary management methods, the traditional resource users are beginning to play a more significant role in the proper utilization of marine resources in the region. This is because most of the contemporary resource management methods have features that are similar or identical to traditional management systems (Joeli 1998). This wealth of resource management knowledge is currently being supported (by organizations like the World Wide Fund for Nature) to complement and improve contemporary management systems. In fact, if intergenerational equity is emphasized, then the wisdom and knowledge accumulated through time should be the basis of the changes that would be made to contemporary management methods. This will enhance the proper management of marine resources (Ban et al. 2019). A strong network currently exists to provide a forum for communities, traditional leaders, government representatives, non-government organizations, researchers and others to share experiences and information.

Rainwater harvesting in India

Rainwater harvesting (RWH) is a simple method by which rainfall is collected for future usage. The collected rainwater may be stored, utilized in different ways or directly used for recharge purposes. With depleting groundwater levels and fluctuating climate conditions, RWH can go a long way to help mitigate these effects. Capturing the rainwater can help recharge local aquifers, reduce urban flooding and most importantly ensure water availability in water-scarce zones. Though the term seems to have picked up greater visibility in the last few years, it was, and is even today, a traditional practice followed in rural India. Some ancient rainwater harvesting methods followed in India include *madakas, aharpynes, surangas, taankas* and many more (Berkes et al. 2000).

The Indian subcontinent has a huge variety of impactful and promising rainwater harvesting practices that extend from dry and wet to arid climatic regions. Presently, the widespread application of RWH is lacking for reasons such as reduced incentives, old colonial-era policies, rise in urbanization and groundwater extraction, and huge irrigation projects for producing cereal. However, the interest in RWH techniques within a holistic water resource management is growing because of increased pressure on natural resources due to population increases (Agarwal & Narain 1997; Ferrand & Cecunjain 2014).

Rice-fish co-culture in China

Rice-fish co-culture, a farming technique for over 1,200 years in south China, was recently designated a 'globally-important agricultural heritage system' by the UN Food and Agriculture Organization. A mutually beneficial relationship has been documented: fish reduce rice pests; rice moderates the fishes' environment, a relationship that reduces by 68% the need for pesticides and by 24% the need for chemical fertilizer compared with monocultures. The findings suggest modern agricultural systems might be improved by exploiting other synergies between species (FAO 2016).

Rice-fish culture (RFC) is the simultaneous or alternate production of fish in a rice field. It consists of stocking the rice field with fish of selected size and species to obtain a fish crop in addition to rice, which is the main crop. If properly implemented it could increase rice farmer's production and income derived from rice and fish. This practice describes the benefits of implementing the fish-rice culture system, gives details on how to prepare the infrastructure for such a system and how to develop the system in a pond, additionally to potential threats like predators (Ibid).

RFC has great potential for being developed in countries with vast areas of irrigated rice fields. Standing water in rice fields promotes the development of a teeming ecosystem with aquatic animals such as fish, ducks, freshwater prawns, marine shrimps, crayfish, crabs, turtle, bivalve, frogs, snails and even insects.

Although cultivating fish reduces the area available for planting rice, the integrated system leads to higher returns than those of rice monoculture. This is due to the fact that within the RFC system, the rice yield is higher, the farmers receive an additional income from the fish sales and in most of the cases, they save on fertilizers and pesticides. In addition, rice-fish farming presents a more efficient use of water. However, it is not recommended for areas with limited water supply as it requires about 26% more water than rice monoculture (De la Cruz et al. 1992). Reports from China, Indonesia and the Philippines indicate that rice-fish farmers' spending on fertilizer is lower.

The benefits of the integrated system are related to nutrition quality and to the interactions between fish and rice systems (De la Cruz et al. 1992). The combination of different plant and animal species makes rice-fish systems nutritionally rich; the rice-fish agroecosystem supplies a wide range of micronutrients, proteins and essential fatty acids that are especially important in the diets of pregnant women and young children; while rice is the main dietary source of carbohydrates, the fish supplies protein, being an important source of cheap and easily digestible animal protein.

In China, by using fish for integrated pest management, rice-fish systems achieve yields comparable to, or even higher than, rice monoculture, while using up to 68% less pesticides. Weed control is generally easier in rice-fish systems because the water levels are higher than in rice-only fields (De la Cruz et al. 1992).

Studies in China found that the presence of rice stem-borers was around 50% less in rice-fish fields. As examples, a single common carp can consume up to 1,000 juvenile golden apple snails every day. The grass carp feeds on a fungus that causes sheath and culm blight. Rice-fish farming is practiced in many countries in the world, particularly in Asia. While each country has evolved its own unique approach and procedures, there are also similarities, common practices and common problems.

Australia

Indigenous fire management in Australia

Indigenous fire management techniques developed thousands of years ago, which today protect large landscapes in Australia, Indonesia, Japan and Venezuela. Early dry season controlled burns create patchy mosaics of burnt country, minimizing destructive late dry season wildfires and maximizing biodiversity protection. In Australia, such projects also create credits sold in carbon markets that support traditional livelihoods.

Sustaining indigenous fire management through carbon markets in Australian indigenous peoples have historically employed customary burning practise to manage the savannah regions. In many cases these practices have ceased, resulting in hot and uncontrolled wildfires late in the annual dry season. Experience in northern Australia shows that strategic reintroduction of traditional patchwork burning early in the dry season can limit the scale and intensity of late dry season fires, reducing emissions of the greenhouse gases that contribute to climate change all while allowing indigenous peoples to generate sustainable livelihoods through the Australian carbon market (Fuller 2020).

In north-western Australia's sparsely populated Kimberley area, aboriginal people have been undertaking traditional fire management for thousands of years, according to the Kimberley Land Council. It says that with the onset of colonization and the removal of the aboriginals from their lands, traditional burning was stopped during the 20th century, which led to the emergence of large, uncontrolled wildfires, usually occurring late in the dry season and destroying important ecosystems and habitats. Now, in the last 25 years, these traditional methods are seeing a reinvigoration with the recognition that western fire prevention methods have not been working.

Some of the indigenous fire management techniques include lighting 'cool' fires in targeted areas during the early dry season between March and

July. This method burns the fuel for larger fires later in the dry season when the weather is very hot and at the same time it also protects habitat for mammals, reptiles, insects and birds.

According to a report in *The New York Times*, over the past decade fire-prevention programs, which are undertaken mainly on aboriginal lands in northern Australia, have 'cut destructive wildfires in half'. Even so, these traditional techniques are not fool-proof methods and need to be used in unison with other fire-prevention techniques (Hill et al. 2013).

Europe

Wilderness area management in Finland

In order to enhance the implementation of Article 8j of the Convention on Biological Diversity (CBD) to better safeguard the rights of the indigenous Saami people and to secure participation and involvement of the Saami people in the protected area management planning process, the application of the voluntary Akwé: Kon Guidelines of the CBD were piloted in the development of the Management Plan of Hammastunturi Wilderness Area in Finnish Lapland. In Finland, the Akwé: Kon Guidelines are meant to be applied in the assessment of cultural, environmental and social impacts of projects and plans which are implemented in the Saami Homeland and may influence the Saami culture, livelihoods and cultural heritage. As a result of the pilot work, a permanent Akwé: Kon model was developed by Metsähallitus, as the administrative authority of state-owned lands, and the Saami Parliament, as the representative of the Saami people, to be used in protected area management and natural resource planning processes in the Homeland of the Saami. The model strengthens the involvement of the Saami people in the management of protected areas and helps the protected area administrative authority to safeguard better the Saami culture and traditional knowledge and livelihoods (Inkeri et al 2019).

The Akwé: Kon Guidelines proved to be applicable to management planning for a protected area in the Saami Homeland in Finnish Lapland. The method was interactive and brought about new ideas and models for practices and procedures. The Akwé: Kon working group provided valuable information about the current situation in the Hammastunturi Wilderness Area, for which the management plan was being developed, from those who use the area. Through the working group, a broader group of people using the area was involved in drawing up the plan. The group also helped in recognizing potential problems as well as values and threats from the point of view of Saami cultural practices. The impact assessment was made into an integrated part of the planning instead of remaining a separate stage after the planning, facilitating revisions to the plan in the drafting stage. Regarding the impacts on Saami culture, the assessment was carried out by the people who use the area, which improved the reliability of the impact assessment (Convention on Biological Diversity 2004).

Indigenous fishing systems in Finland

The Skolt Sami people of Finland participated in a study which adopted indicators of environmental changes based on TEK. The Sami have seen and documented a decline in salmon in the Näätämö River, for instance. Now, based on their knowledge, they are adapting – reducing the number of seine nets they use to catch fish, restoring spawning sites and also taking more pike, which prey on young salmon, as part of their catch. An ongoing project – Snowchange Cooperative – is part of a co-management process between the Sami and the government of Finland.

The project has also gathered information from the Sami about insects, which are temperature dependent and provide an important indicator of a changing Arctic. The Sami have witnessed dramatic changes in the range of insects that are making their way north. The scarbaeid beetle, for example, was documented by Sami people as the invader arrived in the forests of Finland and Norway, far north of its customary range. It has also become part of the Sami oral history.

Tapping into this wisdom is playing an outsized role in sparsely settled places such as the Arctic, where change is happening rapidly – warming is occurring twice as fast as other parts of the world. Tero Mustonen, a Finnish researcher and chief of his village of Selkie, is pioneering the blending of TEK and mainstream science as the director of the Snowchange Cooperative project. 'Remote sensing can detect changes', he says. 'But what happens as a result, what does it mean?' That's where traditional knowledge can come into play as native people who make a living on the landscape as hunters and fishers note the dramatic changes taking place in remote locales – everything from thawing permafrost to change in reindeer migration and other types of biodiversity redistribution.

As part of the pioneering Näätämö River Co-Management Initiative, the Skolts have developed locally devised indicators of environmental change that are more sensitive than Finland's nationally mandated regulatory parameters. Based on traditional knowledge transmitted and updated down the generations, these indicator systems are helping the Skolts and partner scientists to detect and address ecological changes in a pre-emptive fashion.

The Skolt Saami have already identified and begun to restore key salmon spawning sites, helping the embattled fish to reproduce. They have also begun to adapt their own fishing practices and encourage other local subsistence fishers to do the same. Changes include using one net during the salmon season, rather than three, and shifting fishing practices to focus on other species that inhabit the river, including fish that prey on young salmon.

The innovative co-management structure of the project is giving the Skolt Saami a louder voice in matters concerning the Näätämö River, opening spaces for them to share observations and recommendations based on traditional knowledge. 'Co-managers' in the process include scientists and local authorities, brought together in a structure first pioneered in North Karelia, Finland.

North America

Forest management in Wisconsin

Indigenous perspectives and traditions greatly influence forest management on the Menominee Indian Reservation in Wisconsin. Many acknowledge that sustainable forestry, which includes long rotations, single tree selection and long-term monitoring, originated on the Menominee Indian Reservation. Legislation over 100 years ago allowed the tribe to self-govern their forest management practices. Some of the Menominee forest management policies that are driven by cultural mandates include the following: an appropriate harvest rate, use of selection harvests, establishment of a diverse and ample growing stock, long-term monitoring requirements and maintenance of forest goals over industrial goals. Today the Menominee forest stands as an internationally recognized flagship sustainable tribal forestry program and it provides employment for many tribal community members.

Until the era of self-determination from 1972 to the present, few Indian tribes in the United States were able to influence forest management on their reservations. The Menominee Tribe of Wisconsin is a major exception; based upon legislation in 1908, they were able to prevail on the federal government to implement many ideas that are now popular as part of sustainable forest management: long rotation ages, selection harvest practices and long-term monitoring. They also have maintained a mill throughout to support tribal employment. Other tribes have been able to implement their own ideas as their control of reservations has increased; the Intertribal Timber Council has an annual symposium at which tribes exchange ideas about forest management (Trosper 2007).

Essentially, for indigenous ideas to impact forest policy, indigenous people need to have leverage in their dealings with government officials. Although their leverage was not always great, the Menominee affected forest management on their reservation during each of the major periods of

Federal Indian policy. The Menominee Restoration Act in 1972 signalled the implementation of self-determination on Indian reservations in the United States. After self-determination became federal policy, other tribes were able to implement their ideas in forest management and the story of indigenous influence expands to include many sources (Menominee Tribal Enterprises 2013).

Preservation of biodiversity in Mesoamerica

The Maya people of Mesoamerica have much to teach us about farming, experts say. Researchers have found that they preserve an astonishing amount of biodiversity in their forest gardens, in harmony with the surrounding forest. The active gardens found around Maya forest villagers' houses show that it's the most diverse domestic system in the world, integrated into the forest ecosystem.

Ancient Maya people were clever and hardworking farmers who used a variety of techniques to raise enough food to feed the large populations in Maya cities. Their sophistication can be compared to other ancient empires such as the Egyptians. Corn, or maize, was the main staple crop. Maize was grown together with beans and squash as each of the three provide support to the others. Recently, archaeologists also discovered that the Maya grew manioc or cassava, a root that provides a significant amount of carbohydrate in the diet. This discovery solves a longstanding mystery of how the Maya could produce enough nutritious food to feed everyone, considering the land they inhabited and worked with no metal tools or draft animals. In order to deal with rainforest, swampy areas and mountainous hillsides, the Maya had to engineer a variety of Mayan farming methods.

Archaeologists thought for decades that Maya people used slash and burn agriculture, a farming method where trees and other plants are first cut down, then the entire area to be planted is burned. The Maya would then plant in the rich ash that resulted. However, after two or three years, the soil and ash was depleted and must be allowed to lie fallow for five to 15 years. The Maya would then move on to a new area and repeat the process. Some archaeologists realized that the slash and burn technique alone could not have fed the large populations of the Classic era. These experts began to look for other methods the Maya might have used as well as shifting, swidden agriculture (Rank 2019).

Aerial photography provides evidence of raised beds alongside canals. Like the Aztecs, the Maya also farmed field raised up from the bajos, or low, swampy areas. They created these fertile farm areas by digging up the mud from the bottom and placing it on mats made of woven reeds two feet above the water level. In the canals between the beds were fish, turtles and other aquatic life. Water lilies grew in the water and prevented the water from drying up. Raised bed farming was quite labour-intensive but very productive. Each field provided two or three crops a year.

In mountainous areas, the Maya made terraces on the steep hillsides. Small fields are cut into a hillside and held with a retaining wall. These create a series of steps that reduce water runoff and erosion and can be planted with maize or other crops. These terraces make the most productive use of mountainous or hilly land. Here too, the Maya used canals to irrigate the crops.

Besides the three agricultural methods outlined above, the Maya also used forest gardening, planting trees that provided economic benefit for them as food or firewood. Cacao and gum trees were encouraged to grow, for example. The Maya also harvested from the wild, finding tubers, roots and berries they could eat (Rank 2019).

Assessment of endangered species in Canada

Canadian government has begun to incorporate TEK to assess species at risk. Elders of the Heiltsuk First Nation in B.C., for instance, recognized two types of wolves – coastal and inland – previously undocumented by Western scientific methods. With such proven value in only a few examples, imagine how TEK can further inform science.

Still on the patterns of biodiversity, a study in Canada explored traditional knowledge and scientific results about the spawning and migration patterns of fish in the Slave River and Delta. The study showed this dual knowledge system approach to elucidate the broader connectivity of local study regions and how they can improve monitoring programmes by extending beyond the usual context/confines of the present or recent past, as well as increase the spatial and temporal range of system information (Baldwin et al. 2017).

According to the study, blending TEK and Mainstream Science (MS) improves overall understanding of spawning and migration patterns of fish in the Strathcona Reional District (SRD) region. Synergizing TK and MS aids in watershed management in several ways, including: (1) provision of baseline data for assessing environmental change; (2) identification of priority areas for protection or possible remediation; (3) information to support the protection of sensitive species under increased pressure; (4) expanded boundaries and indicators for cumulative effects monitoring; and (5) opportunities for the voices of those living downstream of anthropogenic activities to be heard through the sharing of local knowledge (Ibid).

South America

Traditional medicinal practice in Brazil

Brazil has a long tradition of popular medicine in its different geographical areas (Amazonas, highlands and coast) and different cultural groups (Aruak, Tupi, Guarani).

The promotion and preservation of traditional medicinal knowledge in Brazil Articulação Pacari brings together 47 traditional pharmacies and community-based organizations to cultivate medicinal plants, preserve traditional ecological knowledge and health traditions, and protect biodiversity in Brazil's Cerrado (savannah) biome. In the absence of any comprehensive legislation that legally recognizes traditional health practices, the network has mobilized medicinal plant producers and local health practitioners to create self-regulating policies. Standards have been put in place on the amount of plant used in the preparation of traditional medicines, safety and sanitary conditions for plant processing and sustainable harvesting techniques. In 2012, Articulação Pacari were recipients of the United Nations Development Programme (UNDP)-managed Equator Prize (Zank and Hanazaki 2017).

Recently the documentation of plants used in traditional medicine has been organized in Maceió, Brazil, but pharmacological investigations of these plants have been limited so far. High-level chemical research on natural products is developing quickly in the Federal Universities, but it is only seldom centred on medicinal plants. An exception is constituted by the Institute of Antibiotics of the University of Pernambuco where a multi-year programme on the chemistry of plants with pharmacological properties has been developed. Much interest is being shown in these studies, and the results have potential to be coordinated and developed further (De Mello 1980).

Conclusion

Many indigenous peoples and local communities have lived in sustainable ways for millenniums. Many have knowledge and practices that can maintain and even increase biodiversity on their traditional territories. Furthermore, their knowledge and practices are useful for ecosystem health and maintenance of ecosystem services, from which broader humanity benefits. Often territorial and locally based and possessing an encyclopaedic knowledge of their local environment and its biodiversity, indigenous peoples and local communities are often best placed to economically and optimally manage the local ecosystem, including protected areas. Indigenous peoples and local communities have much to contribute to global discussions concerning sustainability and have a right to participate in matters that may affect them. As proponents and practitioners of both biological and cultural diversity or biocultural diversity, indigenous peoples and local communities have unique insights into possible solutions both locally and globally.

TEK has been used in parallel with Western scientific approaches to characterize and manage resources and ecosystem health. Given the place-based nature and intergenerational transmission of local knowledge inherent in TEK, it can be viewed as a tribal form of citizen science and as a grassroots response to environmental health risks. It can serve as a culturally based framework that allows local community members to actively participate in identifying and addressing lifelong and multiple environmental exposures that affect their health.

Knowledge gained from a TEK perspective, situated in the cultural and spiritual context in which it was acquired, has the potential to improve scientific models of ecosystem and human health and to inform policy and decision-making in important ways. Through collaborative approaches to scientific research and shared policy and resource management decision-making, tribes and the federal government will be better informed to develop models for sustainable practice and to create lasting policies that enhance the health and quality of life for their citizenry.

TEK stands on its own as the indigenous complement to mainstream scientific understanding of environmental and health disparities among Africans, Asians, American Indians, Australians and Europeans. The inclusion of TEK, as well as the broader topic of indigenous knowledge (IK), with mainstream scientific research will contribute to more meaningful and generalizable outcomes. Indigenous knowledge is used as the basis for local-level decision-making in many rural communities. However, conceptual models are needed that discuss, articulate and operationalize IK and TEK principles in relation to studies that explore the environmental, cultural and social determinants of health. Use of these conceptual models and the principles of IK and TEK in research would also require a paradigm shift in how Western-trained scientists understand and respect the 'ways of knowing' shared by indigenous community partners engaged in scientific research.

It is the proposal and operationalization of such conceptual models that the next chapter will address.

References

Agarwal A and Narain S, *Dying Wisdom: Rise, Fall and Potential of India's Traditional Water Harvesting Systems* (Centre for Science and Environment, New Delhi, 1997).

Akpinar Ferrand E and Cecunjanin F, 'Potential of Rainwater Harvesting in a Thirsty World: A Survey of Ancient and Traditional Rainwater Harvesting Applications' [2014] 6(8) *Geography Compass* 395–413.

Baldwin C, Bradford L, Carr M, Doig L, Jardine T, Jones P, Bharadwaj L and Lindenschmidt K, 'Ecological Patterns of Fish Distribution in the Slave River Delta Region, Northwest Territories, Canada, as Relayed by Traditional Knowledge and Mainstream Science' [2017] 34(2) *International Journal of Water Resources Development*, 1–20.

Ban N, E Wilson and Neasloss D, 'Strong Historical and Ongoing Indigenous Marine Governance in the Northeast Pacific Ocean: A Case Study of the Kitasoo/Xai'xais First Nation' [2019] 24(4) *Ecology and Society* 10–21.

Berkes F, Colding J and Folke C, 'Rediscovery of Traditional Ecological Knowledge as Adaptive Management' [2000] 10(5) *Ecological Applications* 1251–1262

Bowen-Jones E and Entwistle A, 'Identifying Appropriate Flagship Species: The Importance of Culture and Local Contexts' [2002] 36(2) *Oryx* 189–195.

Convention on Biological Diversity, *Akwe: Kon Guidelines* (Secretariat of the Convention on Biodiversity, Quebec, 2004).

Convention on Biological Diversity (CBD), *Zero Draft of the Post-2020 Biodiversity Framework* (Secretariat of the Convention on Biodiversity, Quebec, 2004).

Danielsen F, Burgess ND and Balmford A, 'Monitoring Matters: Examining the Potential of Locally Based Approaches' [2005] 14(11) *Biodiversity and Conservation* 2507–2542.

De Mello JF, 'Plants in Traditional Medicine in Brazil' [1980] 2(1) *Journal of Ethnopharmacology* 49–55.

Drew JA, 'Use of Traditional Ecological Knowledge in Marine Conservation' [2005] 19(4) *Conservation Biology* 1286–1293.

FAO, *Growing Rice and Fish – Together: A Chinese Tradition for 1000 Years* (FAO, Rome, 2016).

Fuller T, 'Reducing Fire and Cutting Down Emissions the Aboriginal Way' *New York Times* (16 January 2020).

Hill R, Pert PL, Davies J, Robinson CJ, Walsh F and Falco-Mammone F, 'Indigenous Land Management in Australia: Extent, Scope, Diversity, Barriers and Success Factors' [2013] 6–7 *Cairns: CSIRO Ecosystem Sciences* 14–15.

Inkeri M, Minna T and Sini K, 'Traditional and Local Knowledge in Land use Planning: Insights into the use of the Akwe: Kon Guidelines in Eanodat, Finnish Sapmi' [2019] 24(1) *Ecology and Society* 20–42.

International Institute for Sustainable Development, 'Summary of the Second Session of the Plenary of the Intergovernmental Science-Policy Platform on Biodiversity and Ecosystem Services' [2013] 31(13) *Earth Negotiations Bulletin* 1–14.

Joeli, V, 'Traditional marine resource management practices used in the Pacific Islands: An agenda for change' [1998] 17 *Journal of Ocean and Coastal Management*, 123–136.

Jupiter SD, Cohen PJ, Weeks R, Tawake A and Govan H, 'Locally-Managed Marine Areas: Multiple Objectives and Diverse Strategies' [2014] 20 *Pacific Conservation Biology* 165–179.

Koesoemadinata S and Costa-Pierce BA, 'Development of Rice-Fish Farming in Indonesia: Past, Present and Future' in CR De la Cruz, C Lightfoot, BA Costa-Pierce, VR Carangal and MP Bimbao (eds), *Rice-Fish Research and Development in Asia* (ICLARM Conf. Proc. 24, 1992) 457.

Mahoo H, Mbungu W, Yonah I, Recha J, Radeny M, Kimeli P and Kinyangi J, *Integrating Indigenous Knowledge with Scientific Seasonal Forecasts for Climate Risk Management in Lushoto District in Tanzania* (Climate Change, Agriculture and Food Security (CCAFS) Working Paper, no 103, CGIAR Research Program on CCAFS, Copenhagen, 2015).

Menominee Tribal Enterprises (MTE) Menominee Forestry Centre, *Environmental Assessment for the Menominee Forest Management Plan 2012–2027* (Menominee Tribal Enterprises, Wisconsin, November 2013).

Pierotti R and D Wildcat, 'Traditional Ecological Knowledge: The Third Alternative' [2000] 10(5) *Ecological Applications* 1333–1340.

Rist L, Shaanker R, Milner-Gulland E and Ghazoul J, 'The use of Traditional Ecological Knowledge in Forest Management: An Example from India' [2010] 15(1) *Ecology and Society* 3–24.

Roth G, 'The Position of Farmers' Local Knowledge Within Agricultural Extension, Research and Development Cooperation' [2001] 9(3) *Indigenous Knowledge and Development Monitor* 10–12.

Rank S, 'The Romans-Food' *History on the Net*, (Salem Media Group, Californis 2019).

Schwartzman S, Nepstad D and Moreira A, 'Arguing Tropical Forest Conservation: People Versus Parks' [2000] 14(5) *Conservation Biology* 1370–1374.

Sheil D and A Lawrence, 'Tropical Biologists, Local People and Conservation: New Opportunities for Collaboration' [2004] 19(12) *Trends in Ecology and Evolution* 634–638.

Tchale H, Kumwenda I, Wobst P and Mduna J, *Technical Efficiency of Smallholder Farmers in Malawi: Which Policies Matter Most?* (Centre for Development Research, University of Bonn, Germany, 2005).

Trosper R, 'Indigenous Influence on Forest Management on the Menominee Indian Reservation' [2007] 249 *Forest Ecology and Management* 134–139.

Wehi PM, 'Indigenous Ancestral Sayings Contribute to Modern Conservation Partnerships: Examples Using Phormium Tenax' [2009] 19(1) *Ecological Applications* 267–275.

Zank S, Hanazaki N 'The coexistence of traditional medicine and biomedicine: A study with local health experts in two Brazilian regions' [2017] 12(4) *PLoS One*, 1–17.

5 Synergizing TEK and mainstream science to promote planetary health

> Successful conservation efforts are linked to strong engagement with stakeholders including communities that retain traditional ecological knowledge. Such engagement would comprise assessing problems and proffering solutions.

Introduction

Mainstream science (MS) operates by taking things apart and analyzing the pieces. This reductive process has produced enormously important technological and medical advances. Because of the scientific method, mainstream science appears able to control the environment and provide greater human comfort, and its successes sometimes make it appear infallible.

Native science (traditional science as practised by indigenous peoples) operates by observing the whole and the interaction of the parts. This method, also an organized belief system, has sustained Native peoples and cultures for millennia against nearly overwhelming odds. But, because of this world view, traditional peoples often find themselves ill-prepared to protect their own best interest.

Mainstream science separates the human from the environment and then studies the parts separately as if they had little to do with one another. Mainstream scientific tradition is one in which each piece can be isolated and separated from the larger context. The model is mechanistic and the human runs it (or thinks he does) in the same manner that an engineer operates a train.

Native or indigenous science recognizes the essential interdependence of all of the disparate elements – air, water, land, species of plants and animals, *and* the human. However, because humans have no independent existence "'free' of sunlight, water, microbes in the soil, pollinating insects, etc., they do not reflexively receive absolute priority in Native thinking.

From this indigenous perspective, a culture develops with interdependence ingrained in all its practices and beliefs. Native science believes that the world is sacred and humans are stewards, not masters.

As shown in Chapters Two, Three and Four, traditional ecological knowledge of indigenous peoples is essential in designing and implementing solutions for ecosystems. Traditional knowledge and heritage can contribute to environmental assessments and sustainable ecosystem management. For example, the sustainable production and consumption of indigenous and traditional food has invaluable benefits for natural resources and ecosystems, contributes to a sustainable and healthier diet, and helps mitigate climate change.

This chapter will discuss blueprints for upscaling existing efforts to synergize TEK and mainstream science for more effective environmental conservation strategies that would promote planetary health.

Key international decisions and institutions related to traditional ecological knowledge

The United Nations Environment Programme (UNEP) is working with the UN Permanent Forum on Indigenous Issues to publish work on traditional knowledge for ecosystems restoration and resilience in view of the UN Decade for Ecosystems Restoration (2021–2030).

Key intergovernmental decisions related to traditional knowledge and the environment include the text of the Convention on Biological Diversity (1992) Article 8(j):

> respect, preserve and maintain knowledge, innovations and practices of indigenous and local communities embodying traditional lifestyles relevant for the conservation and sustainable use of biological diversity and promote their wider application with the approval and involvement of the holders of such knowledge, innovations and practices and encourage the equitable sharing of the benefits arising from the utilization of such knowledge innovations and practices;

Also, article 10(c) provides as follows: 'Protect and encourage customary use of biological resources in accordance with traditional cultural practices that are compatible with conservation or sustainable use requirements'. Article 17.2 goes ahead to provide that 'exchange of information' includes specialized knowledge, indigenous and traditional knowledge as well as repatriation of information (Task 15 of the 8(j) programme of work). Article 18.4 encourages the development of methods of cooperation for the use of technologies, including traditional technologies.

Also, under the Strategic Plan for Biodiversity 2011–2020 (adopted by the United Nations General Assembly), by 2020 the traditional knowledge, innovations and practices of indigenous and local communities relevant for the conservation and sustainable use of biodiversity, and their customary use of biological resources, are respected, subject to national legislation and relevant international obligations, and fully integrated and reflected in the implementation of the Convention with the full and effective participation of indigenous and local communities, at all relevant levels.[1] Target 11 of the document provides that by 2020, at least 17% of terrestrial and inland water, and 10% of coastal and marine areas, especially areas of particular importance for biodiversity and ecosystem services, are conserved through effectively and equitably managed, ecologically representative and well-connected systems of protected areas and other effective area-based conservation measures, and integrated into the wider landscapes and seascapes.

At the 2nd plenary session of the Intergovernmental Platform on Biodiversity and Ecosystem Services (IPBES) in Antalya in 2013, delegates considered recommendations from an international workshop on traditional knowledge held in Tokyo in June. The outcome report emphasized that the IPBES conceptual framework must accommodate indigenous and local knowledge and world views in an appropriate, respectful manner. The expert group emphasized that indigenous peoples' and communities' conceptualization of relationships between life's ecological, social and spiritual spheres is reflected throughout their management and knowledge systems (IISD 2013).

These, according to the report, ought to complement science-based representations and form an integral part of the IPBES conceptual framework through a meaningful and active engagement in all relevant aspects of its work and across all of its functions. Gaps in knowledge must be identified and capacity built for the interface between policy and knowledge – in all its forms. That means developing a process through which scientific and policy communities recognize, consider and build synergies with indigenous and local knowledge in the conservation and sustainable use of biodiversity and ecosystem services. Both the Convention on Biological Diversity (CBD) and the IPBES are important arenas for enabling such dialogue across knowledge systems. Both recognize and respect traditional knowledge, innovation and practices (CBD Article 8(j)) and its experiences even though there are differences in terminology – where CBD talks about 'traditional knowledge', IPBES has chosen the term 'indigenous and local knowledge'.

The approach of the CBD, when it was agreed in 1992, broke new ground for understanding the importance of traditional knowledge related to biodiversity resources. Since IPBES was established over 20 years after the CBD, with the ambition of treating knowledge systems equally in its assessments, the insights about the value of indigenous and local knowledge have had increased attention. Within the CBD, the International Indigenous Forum on Biodiversity (IIFB) was formed in 1996, and serves as the caucus where all indigenous peoples and local communities (IPLCs) come together during CBD meetings. The process of building these institutions was useful when it came to setting up a similar caucus within the IPBES, the International Indigenous Forum on Biodiversity and Ecosystem Services (IIF BES). This is the group of IPLC participants present and gathering at IPBES Plenary meetings.

Chapter Four discussed sporadic applications of TEK to conservation policies, showing its potential to make such policies more holistic and effective. However, it must be quickly noted here that combining traditional ecological knowledge and mainstream science can be a challenge, and there are dangers involved that require consideration. Menzies and Butler provide a cautionary note that the 'danger of TEK research is that it can simply make TEK a tool of WS, rather than a complementary approach to resource management' (Menzies & Butler 2006). Hence, data sensitivity and proprietary rights of traditional ecological knowledge are critically important to establish. The best method is to have the indigenous community decide who is able to utilize and speak for different aspects of TEK. Further, any attempt endeavouring to integrate traditional knowledge for biodiversity conservation and sustainability of natural resources should be based on the principle that traditional knowledge often cannot be dissociated from its cultural and institutional setting.

The following sections will review literature on interfacing formal and traditional institutions and knowledge systems, and further discusses blueprints on effective strategies for synergizing both sciences with a view to promoting the formulation of holistic environmental protection policies within the context of threatening pandemics.

Literature review

Conservation potentials of interfacing formal and informal (traditional) institutions

Formal institutions are created, communicated and enforced through channels of generally accepted official organizations (courts, legislatures, bureaucracies) and state-enforced rules (constitutions, laws, regulations) (Leach et al. 1997). They are articulated by laws, written contracts and other codified practices. They make important contributions in the implementation of strategies and technologies of natural resource management and do so with a degree of appropriateness and legitimacy in the eyes of community members. Formal institutions are suitable for the implementation of natural resources management strategies because of their ability to build on the existing bureaucratic structures and the authority vested in state organizations (Shyamsundar et al. 2005).

Informal institutions, on the other hand, are traditional governance arrangements including chieftaincy and traditional priesthood systems and cultural belief systems. It has been observed that where traditional leadership is strong and legitimate, their influence had a corresponding sustainable impact on environmental resources (Larcom et al. 2016; Shackleton et al. 2002). One unique trait associated with traditional institutions and environmental conservation is the animate role attached to the resources which make their protection more like a ritual and linked to the people's well-being (Larcom et al. 2016). Natural resources in the TEK context have not only been regarded for the products and valuable ecological services derived from them, but those resources such as trees, animals, water bodies and aquatic lives, and mountains have been the linchpin of the people's religion and cultural beliefs and were to be kept free from abuse by human activities (Shackleton et al. 2002).

Informal institutions from this perspective thus connote customary rights or pre-existing rules passed down from generation to generation, ostensibly to protect, maintain and sustain natural resources within a particular context mostly not codified into law (Otsuka & Place 2002). They are promulgated, monitored, enforced and sustained within the culture and narratives of a given community, albeit, they may appear questionable to people from a different culture or context (Yeboah-Assiamah et al. 2017; Ali 2016).

As much as formal institutions may promote the conservation of biodiversity and other natural resources, they can also co-exist with informal institutions. The fact that people at the local level know each other better, have more rapport and sense of belonging, provides opportunities for cooperation and collective action for managing natural resources on a self-ruling and self-sufficient basis. In rural areas, local institutions usually include councils of elders, traditional midwives and rainmakers, as well as some spiritually significant parts of the landscape, such as sacred forests and trees. Sacredness bestowed on some trees or forests has been found to reflect important ecological functions and to protect public goods and environmental well-being (Martin et al. 2016; Meliyo et al. 2006;). Such trees or forests are, therefore, traditionally protected through norms and regulations, while breaking these rules might attract severe punishment from spirits (Martin et al. 2016; Laurell & Åke 2000). Apart from being the warehouse for indigenous knowledge and beliefs, local institutions have the potential to effectively link service providers and the local communities. Local institutions surround and connect communities and interact with other institutional systems, such as the local government, to articulate community needs. By understanding where local institutions are likely to form and what issues they are best suited to address, state and federal government agencies can better work with local organizations to address the complexities of natural resource management. Because of their dynamic nature, local institutions are more efficient in promoting sustainability than formal policies and laws regarding resource management (Dixon & Wood 2007).

Whereas technologies and land management practices enable the transformation of resources and determine the pace, cost and effectiveness of change, institutions determine whether and how the relationship between technologies, environment and people would be viewed now and in the future. Uphoff (1992) argues that local institutions are more likely to be successful in natural resource management where the resource is 'bounded', that is, known and predictable rather than shifting and variable, and where the users themselves are an identifiable group or community with their own authority structures.

The discourse on environmental conservation institutions involves interplay of rules, norms structures and actors together with their interactions. Institutional analysis involves the task of identifying the possible multiple and overlapping rules, the groups and individuals affected by such rules and the processes by which the particular sets of rules change in a given situation (Agrawal & Gibson 1993). Any treatise aimed at adequately assessing natural resource institutionalism ought to schematically analyze (i) the institutional arrangement; (ii) the nature of the institutional arrangement [whether polycentric or hierarchical]; (iii) it is also imperative to analyze the action arena to determine the extent to which different stakeholders or actors make informed decisions and (iv) the rules and strategies that structure relationships between actors and resources (Yeboah-Assiamah et al. 2017). Thus, in order to balance livelihood and conservation objectives, it is essential to engage local communities in the management of natural resources. For successful engagement of local communities, policymakers need to recognize and work with local institutions (Lenjisa 2015). This is because of their role as custodians of local ecological knowledge (Donnely-Roark, Ouedraogo, and Ye (2001), their ability to mobilize collective action (Gupta 1992; Olate 2003) and their ability to connect members of different communities (Donnely-Roark, Ouedraogo, and Ye (2003), all of which are fundamental to effective natural resource management as discussed in Chapters Two and Four.

Working with multiple knowledge systems

The acquisition of knowledge entails processes of learning, re-framing and understanding (Mistry & Beradi 2016). In this process of negotiation, tensions can arise at the interface between actors with different views of what constitutes reliable or useful knowledge. Those tensions must be managed effectively if the potential benefits of knowledge are to be realized Thus, to achieve the level of cooperation required for effective biodiversity conservation, best-practice knowledge production must be facilitated. This requires a 'knowledge partnership' approach that is focused on collaboration and emphasizes relationships as opposed to difference and/or incongruence (Berkes 2009).

Globally, there is increasing acknowledgement that a majority of the world's biodiversity, threatened species and ecosystem services exist within indigenous estates (Maffi & Woodley 2012; Renwick et al. 2017). There seems to be a direct correlation between linguistic and cultural diversity and biodiversity (Gorenflo et al. 2012; Maffi & Woodley 2012;), which hints at the role different knowledge systems can play in innovation and research. As such, though often contested (Berkes 2008), arguments that indigenous peoples' knowledge and practices play a key role in maintaining globally significant environmental assets are gaining strength (Bohensky et al. 2013; Tengo et al. 2014; Berkes 2015).

As seen in Chapter Four, TEK has the potential to provide useful/useable knowledge, methods, theory and practice to assist in the sustainable management of natural resources, 2014; Tengö et al. 2014). The chapter also demonstrated that there are numerous relevant studies that exemplify knowledge integration practice. However, there is no one knowledge integration process or practice that can be applied universally. Subjectivities related to context, aspirations and the answers being sought determine each process (Reid et al. 2006; Berkes 2008; Danielsen et al. 2009, 2014; Tengö et al. 2014). Further, effective integration is largely dependent on the degree to which partnerships are underpinned by good faith (Nadasdy 1999; Verran 2002, 2008; Christie 2006, 2007; Muller 2014).

It is important to emphasize here that TEK does *not* have all the answers to biodiversity threat. It, however, plays a significant role in the understanding and management of species. Thus, issues such as climate change, invasions of feral animals and weeds, and zoonotic induced pandemics are complex challenges that may require co-produced knowledge and management solutions. Given the resources, indigenous people will in many cases be able to generate solutions to these problems in their locations, though they may find mainstream scientific knowledge and management practices of use. They will also be experienced in using mainstream scientific knowledge methods to build richer knowledge bases about species. Likewise, managers of national parks and private landowners will be able to learn a great deal from the knowledge and experience of traditional owners. The harnessing of all available knowledge widens the scope, depth and value of knowledge that can be used to inform management, conservation and environmental monitoring of biodiversity. There is a strong (erroneous) tendency to refer to TEK in an opaque way as an add-on to mainstream science. It is therefore imperative to clarify the nature of indigenous knowledge and position it as equal to mainstream science so as to realize the full potential of Multiple Knowledge Systems (MKS) or Multiple Evidence Based (MEB) approaches to conserving biodiversity and other natural resources.

MKS and MEB as presented by Tengö et al. (2014) is an approach that can work with diverse knowledge systems to produce an enriched picture of any given phenomenon. It depicts graphically the notion of 'science and other knowledges' being worked together to build a more comprehensive knowledge base than could be achieved by any one knowledge system alone. The purpose is to mobilize diverse and potentially otherwise disparate knowledge to co-generate mutual learning across knowledge systems (Austin et al. 2018).

The incorporation of MKS into integrated assessments of environmental and social status has been established as critical (Pahl-Wostl 2003;). It is crucial to engage MKS in the creation policies aimed at promoting planetary health for several reasons: (1) Well-being of biodiversity implies that the perspectives of and information from local (that is, community or village level) residents and users of the species are important to understand. Local people have particular knowledge about the biodiversity that they live and work within, as well as their own associated well-being, that others do not. (2) Ideally, an assessment at any given scale should meet the needs of resource users and managers at that scale, who should be involved in defining the issues of concern. Thus, local-level management depends on the voices of local people, which are all too often not heard, or else are ignored or misunderstood. (3) The use of multidisciplinary and multistakeholder perspectives is important in order to understand the links between biodiversity and human well-being; an assessment is usually enhanced when informed by a variety of research, scientific or other perspectives. Just as people from different locations speak a different language and express their ideas differently, so do scientists trained in different disciplines and people working in different organizations (NGOs, development agencies, etc.). (4) The gap between research and policy is often recognized but rarely solved; exploring the differences in framing, representing and legitimating knowledge among scientific researchers and policy/decision-makers will help to close this gap. (5) The process of rigorous documentation and use of local and traditional ecological knowledge is often seen as empowering local resource users, as it can link them to decision-making at higher scales and possibly catalyze decision-making capacity at the local level. However, novel institutional arrangements are often necessary for this to occur (World

Resources Institute 2003) because to build social-ecological resilience, knowledge needs to be embedded in an institutional context that enables application and learning from experience over time (Tengo et al. 2017).

MKS has helped frame the work of the Intergovernmental Panel for Biodiversity and Ecosystem Services (IPBES) (for example, see the IPBES global assessment of pollination services and regional assessments.[2] Such efforts have been forged through a commitment by all participants to value diversity (sociocultural-ecological) and creating knowledge pathways and partnerships that link local, regional, national and global policy and management of biodiversity (Tengö et al. 2016). This is not without significant challenges and so barriers to understanding that may arise when knowledge derived from different processes is exchanged have to be overcome. These barriers may take the form of dismissing the arguments of an unfamiliar discipline, questioning the validity of data analysis, or, more simply, finding it too difficult to work within a foreign construct/world view because the logic does not make sense. Possible cognitive barriers have also been identified, such as the absence of shared world views (Hill et al 2015; Berkes 2015; Houde 2007).

The MKS/MEB positions indigenous, local and scientific knowledge systems (among others) as 'different manifestations of valid and useful knowledge that generates complementary evidence for sustainable use of biodiversity' (Tengö et al. 2016). It has a focus of 'letting each knowledge systems speak for itself, within its own context, without assigning one dominant knowledge system with the role of external validator' (Tengo et al. 2014). The outcome can be thought of as knowledge weaving through collaborative pathways, activities and efforts that respects the integrity of each knowledge system (Johnson et al. 2016). In a practical sense, the MKS/MEB approach allows for accurate, efficient identification of gaps in the knowledge base and opportunities for collaborative research engagements.

The fundamental requirement for the MEB to function effectively is empowerment and capacity development of practitioners from all participating knowledge systems (Christie 2006, 2007; Tengo & Malmer 2012; Tengo et al. 2014; Tengo et al. 2017; Austin et al. 2017a). This is based on the presumption that indigenous knowledge holders are able to collectively organize and mobilize their knowledge appropriately at organizational and institutional levels. The articulation of knowledge across cultural boundaries requires conscious effort within the context of relationships of 'good faith' and mutual respect (Mowo 2013).

The main challenge when working with multiple knowledges concerns the fundamentally different natures of TEK and MSK and the partial incompatibility that exists. To highlight this issue, it is useful to look at the three different types of knowledge in IK: *as content*; *as process* and *as beliefs*.

Knowledge as content is the most easily recognizable form of IK from the perspective of MSK. This refers to the knowledge/information held by indigenous peoples that can be easily passed on from one person to the next (Austin et al. 2018; Berkes 2015). An example of this can be found in the abundant knowledge that local people have for local species such as their life cycles, their distributions and the type of habitat in which they live. This form of IK is most likely to be seen as compatible with MSK practitioner perspectives and, as such, is relatively simple to incorporate into collaborative knowledge initiatives.

Traditional ecological knowledge begins at the level of local and empirical knowledge of species and the environment. It proceeds to the level of practice, which requires understanding local ecological processes and how to live and work with them. Practice requires rules-in-use or institutions to guide how a group of people relate to their environment and resources (Berkes 2015). Indigenous knowledge as process is less easily accommodated by MSK. Though similar in terms of being underpinned by a process of curiosity-observation-inference, the way that knowledge is constructed by indigenous knowledge holders is fundamentally different to the scientific method (Berkes 2015), as discussed in Chapter 3.

IK process comprises observing, discussing and making sense of information that are passed on from generation to generation (Berkes 2012b). Indigenous knowledge holders use dynamic, experiential and highly adaptive mechanisms to construct knowledge that incorporates a significant component of learning-by-doing (Berkes 2015).[3] Ways of knowing which things are important enough to be observed, how to observe them and how to make sense of observations are passed on across generations. A useful example for highlighting the importance of IK as process, and the challenge of incorporating it in knowledge integration work, is climate change. Though climate change is a relatively new phenomenon and, as such, indigenous knowledge holders have little historical experience of it, the types of observations made, methods used, new knowledge formulated and implications may be different, though equally legitimate and useful to those of MSK (Berkes 2015). This process of constructing knowledge is place and people specific and, for this reason, is difficult to accept for MSK practitioners who place emphasis on the absolute, universal and objective scientific method (Austin et al. 2018).

The third aspect of IK that poses considerable challenge for integration with MSK is knowledge as belief, which informs, is informed by and is practiced by indigenous knowledge holders.[4] It is this idea that the world view of indigenous peoples shapes human-environment relationships and the types of 'environments' that are imagined and/or observed, that sits most uncomfortably with MSK (Berkes 2015). An example of this difference can be seen in the beliefs passed on through stories and other oral traditions by many indigenous peoples that everything that exists (including animals, plants, water, rocks, etc.) has its own agency (Unuigbe 2020). Humans are not offered a privileged place in most indigenous cosmologies, having no special powers of objectivity, rational thought or special mastery of knowledge. Rather, humans and non-humans cannot be separated and enjoy relationships of reciprocity and mutual respect. This is fundamentally different to MSK's claims that the material world is only knowable through careful and objective measurement that aims to limit the subjectivity inherent in human observations of the natural world. As such, accommodating this fundamental difference in world view is a challenge for collaborative knowledge integration work. While many scientists can determine the implications that knowledge as belief represent, say for the discipline of ecology, for most it is not possible to fully embrace the possibility of alternate perspectives on reality. For example, the ethic of respect discussed in Chapter Two can be understood broadly by MSK practitioners as a holistic approach to conserving biodiversity and other natural resources. However, the concept that performing ceremony and other obligations can have a direct, material influence over the presence or absence of particular species is difficult for many scientists to believe as true (Austin et al. 2018).

Whether expressed publically or not, there is scepticism about the contemporary existence and/or effectiveness of IK (Austin et al. 2018; Nadasdy, 1999). Equally, indigenous people are often sceptical about the motives and intent of scientists who want to 'capture' or 'communicate' their knowledge (Austin et al. 2018; Verran 2008; Christie 2007, 2006; Nadasdy 1999). These are both valid concerns, influencing and being influenced by power relations. Knowledge integration strategies for policy influence must offer solutions to relative power imbalances between local Indigenous peoples' and their partners. Indeed, they must promote egalitarianism and ensure that all parties begin and remain on an equal footing (Wohling 2009; Tengo et al. 2014; Berkes 2015). Ignoring power makes it increasingly likely that attempts at knowledge integration will reinforce rather than break down biases of knowledge systems in the conservation of biodiversity and other natural resources (Sillitoe 1998; Nadasdy 1999). One way of dealing with this mutual scepticism is to employ 'good faith' in recognizing different theoretical, methodological and practical approaches to understanding and interacting with the biophysical world (Verran 2002, 2013; Christie 2006, 2007;; Tengo et al. 2014). The effectiveness of indigenous peoples' knowledge-practices-beliefs should be assessed based on the outcomes of biodiversity conservation activities, in truly postcolonial contexts where diverse knowledge models are accepted as legitimate and useful until proven otherwise. A further challenge is ensuring that indigenous people themselves are able to continue building their capacity to conduct MSK based research and management for biodiversity conservation. Indigenous people should, for instance, have the ability to both generate and gather data and information from both IK and MSK (and other knowledge systems) to build an enriched picture of biodiversity conservation and support best-practice two-way management.

Implications of institutional and knowledge linkages for planetary health

Planetary health challenges such as the COVID-19 pandemic have been largely linked to zoonosis, as discussed in Chapter Two. With a significant linkage of zoonosis to loss of biodiversity, pragmatism on the conservation (or possible restoration) of the later is therefore highly imperative.

Given that research is conclusive on the fact that nature is declining less rapidly on lands that indigenous people manage than in other areas (IPBES 2019), their knowledge systems and institutions must have something significant to contribute to current policies. For emphasis, deforestation rates are significantly lower in places where indigenous people securely hold land. When indigenous peoples have communal control of land, biodiversity loss is noticeably less. This suggests that it is the land-management practices of many indigenous communities that are keeping species numbers high; thus, collaborating with indigenous land stewards will likely be essential in ensuring that species survive and thrive to promote planetary health (UNEP 2020(a); (b)).

The groundbreaking IPBES report on the Summary for Policymakers of the Global Assessment Report on Biodiversity and Ecosystem Services (IPBES 2019) asserts that regional and global scenarios currently lack and would benefit from an explicit consideration of the views, perspectives and rights of indigenous peoples and local communities, their knowledge and understanding of large regions and ecosystems, and their desired future development pathways. Recognition of the knowledge, innovations and practices, institutions and values of indigenous peoples and local communities and their inclusion and participation in environmental governance often enhances their quality of life, as well as nature conservation, restoration and sustainable use. Their positive contributions to sustainability can be facilitated through national recognition of land tenure, access and resource rights in accordance with national legislation, the application of free, prior and informed consent, and improved collaboration, fair and equitable sharing of benefits arising from the use, and co-management arrangements with local communities (Ibid).

Table 5.1 is a representation of selected statistics from the IPBES report on sectors directly related to, or incidental to, biodiversity loss.

Table 5.1 Selected statistics on biodiversity loss

Climate change 100% increase since 1980 in greenhouse gas emissions, raising average global temperature by at least 0.7 degree 5%: estimated fraction of species at risk of extinction from 2°C warming alone, rising to 16% at 4.3°C warming For global warming of 1.5–2 degrees, the majority of terrestrial species ranges are projected to shrink profoundly
Food and agriculture +/-25%: greenhouse gas emissions caused by land clearing, crop production and fertilization, with animal-based food contributing 75% to that figure >75%: global food crop types that rely on animal pollination 5.6 gigatons: annual CO2 emissions sequestered in marine and terrestrial ecosystems – equivalent to 60% of global fossil fuel emission
Forests 50%: agricultural expansion that occurred at the expense of forests 290 million ha (+/-6%): native forest cover lost from 1990–2015 due to clearing and wood harvesting 50%: agricultural expansion that occurred at the expense of forests
Health 17%: infectious diseases spread by animal vectors, causing >700,000 annual deaths
Mining and energy <1%: total land used for mining, but the industry has significant negative impacts on biodiversity, emissions, water quality and human health +/-17,000: large-scale mining sites (in 171 countries), mostly managed by 616 international corporations
Oceans and fishing +/-50%: live coral cover of reefs lost since 1870s 100–300 million: people in coastal areas at increased risk due to loss of coastal habitat protection
Species, populations and varieties of plants and animals Up to 1 million: species threatened with extinction, many within decades >500,000 (+/-9%): share of the world's estimated 5.9 million terrestrial species with insufficient habitat for long-term survival without habitat restoration >40%: amphibian species threatened with extinction 25%: average proportion of species threatened with extinction across terrestrial, freshwater and marine vertebrate, invertebrate and plant groups that have been studied in sufficient detail

At least 680: vertebrate species driven to extinction by human actions since the 16th century

+/-10%: tentative estimate of proportion of insect species threatened with extinction

>20%: decline in average abundance of native species in most major terrestrial biomes, mostly since 1900

+/-560 (+/-10%): domesticated breeds of mammals were extinct by 2016, with at least 1,000 more threatened

30%: reduction in global terrestrial habitat integrity caused by habitat loss and deterioration

47%: proportion of terrestrial flightless mammals and 23% of threatened birds whose distributions may have been negatively impacted by climate change already

>6: species of ungulate (hoofed mammals) would likely be extinct or surviving only in captivity today without conservation measures

Urbanization, development and socioeconomic issues
>100%: growth of urban areas since 1992
>2,500: conflicts over fossil fuels, water, food and land currently occurring worldwide

Despite the differences in scope and focus of local institutions, they are characterized by several commonalities, which can be harnessed for biodiversity conservation. First, their activities are generally related directly or indirectly to conservation. Second, they are generally structured to address either or both environmental and socioeconomic goals. Third, there is the strong influence of traditional leaders that can be used to expand the institutions' sphere of influence in the area. Fourth, they have some form of decision-making and local authority over natural resources biodiversity conservation. Finally, they have deep understanding of people's relations with their environment and draw freely from a rich local ecological knowledge of the area. Essentially, judicious management of biodiversity depends on recognition of local institutions and working with them in conjunction with modern formal institutions.

Also, taking knowledge sharing seriously creates opportunity for innovative environmental management approaches that would mitigate future planetary health threats. The cultural practices of science and IK could be used and adapted to confer normative authority on regional standard-setting in a way that resonates with local people, whose commitment might also ensure compliance with conservation regulations.

Synergizing TEK and mainstream science 67

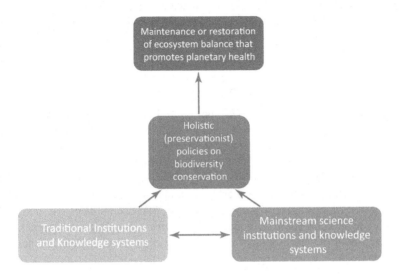

Figure 5.1 Implication of TEK and MSK interface for planetary health.

The foregoing subsections have reiterated the theory of TEK as adaptive environmental management as discussed in Chapter Three as well as case studies on integrated TEK in Chapter Four. The next section discusses how policymakers can be intentional about structurally operationalizing the institutional and knowledge linkages (discussed above) to inform new environmental conservation policies or modify existing ones. The policy recommendations discussed contemplate gender considerations, government support, protection of intellectual property rights, conservation awareness (education) and capacity building.

Policy directions for linking TEK and WS knowledge systems and institutions (Dixon & Wood 2007; Austin et al. 2018)

Integrating biodiversity conservation into health and recreational activities

Health and recreational institutions can be sensitized to include biodiversity conservation in their activities given the reliance of traditional healers on local biodiversity and the reliance of recreational institutions on biodiversity and other natural resources for fundraising. Recreational institutions offer an opportunity to bring individuals with similar interests together. As a means of raising funds to meet costs related to their groups, members of these institutions can be involved in various production activities including cultivation of high-value crops.

Creating conservation awareness

Resource management objectives such as soil conservation and the allocation of irrigation water can be achieved if backed up by appropriate technological, policy and institutional innovations. One way of shoring up traditional institutions is to encourage interaction between them and formal institutions involved in resource management. For example, each village/district/county can have an environmental management committee under the village/district/county government. Such committees can make use of institutions like traditional dance groups in creating awareness on management of natural resources including water sources, forests and soil conservation.

Systemic studies to 'decode' indigenous knowledge

The declining importance of traditional beliefs, rituals, sacred forests and trees is disadvantageous to the management of natural resources. Being unable to provide a 'scientific' basis for such beliefs, guardians of these institutions and traditional forms of management were deemed 'primitive'. Systematic studies directed towards decoding the 'indigenous knowledge' embedded in some useful traditional beliefs are, therefore, necessary to provide the practitioners of local institutions with information by which they can defend some of these beliefs in a more scientific way.

Gender considerations

Women need to be equally and actively involved in processes to conserve and sustainably use biodiversity because they play critical roles as primary land managers and resource users, and they face disproportionate impacts both from biodiversity loss and gender-blind conservation measures. Enabling women's full engagement in biodiversity decisions is critical to ensure that biodiversity conservation and sustainable use efforts are successful in the long term. Without the contributions and buy-in of women and girls, these efforts risk overlooking the root causes of biodiversity loss.

Traditional knowledge reveals that including women in forest and fishery management groups can result in better resource governance and conservation outcomes. Women's particular roles and responsibilities within the household, community and society lead them to develop unique knowledge related to biodiversity, shaped by their specific needs and priorities. They are thereby in a unique position to bring different perspectives and new solutions to addressing biodiversity concerns. Gender barriers in land tenure reforms, for instance, can be removed/reduced as this would lead to a significant increase in soil conservation investment by female-headed households.

Supporting decision-making and planning on conservation

Local governments (as formal institutions) are expected to support and bring traditional informal institutions into decision-making and planning. As seen above, formal institutions constitute an arena for local politics with important local players within it, a structure of opportunities for the negotiation of the distribution of resources and a significant space for stakeholders to negotiate arrangements of natural resources management. Traditional informal institutions have a long history of successful land management and conflict resolution. Traditional leaders are living institutions themselves; their sphere of influence is founded and deeply rooted in the local culture. Therefore, local informal institutions that persist through the advent of modernization should be tapped and supported to align with formal institutions so that local support is generated for effective natural resources management.

Government support

Implementing an interface between formal and traditional institutions requires defined role playing especially by the government. Sustainable success in conservation activities would require the government harmonizing formal and informal institutions on conservation and natural resources management. The sustainability of forest management, for instance, would be secured through informal (traditional) institutions that are protected by state formal rules in the form of forest areas and facilitated by reliable local governments. This type of model gets local legitimacy and succeeds in achieving socioeconomic and environmental objectives. The village/district/county administration as facilitators to bridge the informal and formal institutions is an authority structure allowing the traditional institutions of forest conservation area to be sustained.

The presence, communication, rewards and knowledge provision are the roles of government officials to reinforce the legitimacy of such local institutions. These institutions therefore require external institutions to maintain their legitimacy. This rebuffs views that strengthening community-based forestry practices would tamper the roles of state governments.

Understanding the strengths and weaknesses of the traditional institution of natural forest conservation is an important pillar for forest conservation and development.

Establishing integrated dialogues and knowledge systems

Integrating or co-producing knowledge systems requires collaboratively identifying problems and goals, and effective implementation of agreed action plans. Highlighted below are steps that can be explored in transdisciplinary research projects or assessments to effectively establish multiple knowledge systems:

1. **Establishment of mechanisms for respectful dialogue**

There should be an assessment of differences in experiences, methods, goals, knowledge systems and power relations between all participating partners. Recognition of these differences, and the implications they have for working relationships, is crucial to mobilizing all available knowledge systems.

Having established dialogue and considered capacity to participate as equals, the next task is to mobilize all available knowledge systems (in this context, TEK and MSK) to define the scope of the project/assessment; map knowledge gaps and identify research goals. In MSK-based research, this task is usually conducted by reviewing all available literature on the topic. It is important that MSK practitioners are able to summarize and explain the literature to indigenous colleagues and, thus, the nature of the knowledge gap that they are seeking to fill in a way that is comprehensible in an intercultural context. This may require some translation, not only of languages, but also of concepts, theories and beliefs.

Mobilizing available TEK will usually involve the participation of relevant IK holders, as identified by the local community, and who have access to relevant, useable knowledge on the topic. The local community will be able to identify the right people to participate as knowledge holders and, through this process, validate that the IK being mobilized is accurate and trustworthy. Indigenous partners will also need to be able to clearly articulate their knowledge gaps on the research topic, which requires a certain capacity to communicate effectively between cultures. Likewise, this will likely involve translation of languages, concepts and beliefs.

2. Plan for a holistic picture

A decision will need to be made by the team as to whether the work will use co-production or parallel integration to drive methods and data collection and produce results. If co-production is identified as the best approach, the project team will need to collaboratively identify which methods they will employ to do the work. Here the focus should be on ensuring that working relationships are of an egalitarian nature, that comparative advantages of each knowledge system are leveraged and that each member of the project has the capacity to participate equally.

If parallel integration is identified as the best approach, each of the knowledge holders/producers will be able to identify the most appropriate methods from their respective knowledge systems. However, the project team will need to clearly understand the types of results that will be produced and will need to identify appropriate mechanisms for integrating results.

Project teams may choose to combine the various forms of knowledge to produce integrated reports or, alternatively, the results of each group of knowledge holders/producers can be presented alongside each other. The key here is to ensure that equal legitimacy is afforded to the knowledge produced, irrespective of the methods used to produce it.

3. Implementation of identified activities

TEK holders will need to consider how they can appropriately and safely communicate their knowledge for the specific purpose of the research project/assessment – a challenge of bridging epistemologies which very few TEK holders would have previously encountered. Similarly, many MSK practitioners will be met with new challenges when constructing knowledge. Often cited issues such as dealing with distances, climate, accessibility, costs and working alongside indigenous peoples (which may be new for many) will be faced. Only by employing the guiding principle of good faith, acknowledging respective capacity deficits and power relationships, and ensuring that project budgets are sufficient to meet the requirements of all knowledge systems and practitioners can holistic objectives be met.

4. Analyze, interpret and communicate the data

To realize the most impact from integrated and/or co-produced outcomes, a conscious and good faith approach to examining the similarities, complementarities and/or contradictions of findings need to be discussed. This need not focus on achieving consensus, which can lead to adversarial or bad faith participation, but should focus on assessing what new insights have been produced, how they contribute to the enriched knowledge base and what knowledge gaps remain for further investigation. Again, crucial to this is the role of knowledge brokers who can assist in the translation of results, analysis and interpretations.

For example, the MSK community will need clearly written and academically rigorous reports, journal publications and/or conference presentations to meet their audience. Indigenous partners may prefer products that contain less scientific jargon, more imagery and communicate implications of research for local governance and practice. Using digital media products, such as film, websites and social media, may also be useful for targeting multiple audiences. Irrespective of the nature of research products and communication strategies adopted, the key to ensuring relevance and effectiveness of delivering messages to key audiences will be ensuring that all stakeholders have meaningful roles to play in the production and communication process. Importantly, all outputs should be comprehensible to all of the participants in the research project, which may require patient translation by TEK and MSK practitioners respectively.

5. Need for intercultural knowledge brokerage

Fundamental to ensuring that an MKS/MEB approach is applied effectively is the role of 'knowledge brokers'. Knowledge brokers are people (individuals/organizations, indigenous/non-indigenous) who have the capacity to create meaningful, appropriate and functional linkages and relationships between otherwise disparate knowledge holders/producers. These people or organizations need sufficient self-awareness and understanding of the nature and implications of their own knowledge system, plus sufficient understanding, sympathy and good faith relationships to others and their respective knowledge systems. Ensuring that knowledge brokers are in place will assist in more productive research collaborations that are able to fully mobilize all available knowledge systems for the production of cutting-edge research that enhances decision-making, policy and management actions.

6. Scaling up traditional ecological knowledge

There is a need to make IK matter at larger than local scales while avoiding the loss of legitimacy among knowledge holders as well as decision-makers at different levels. However, given the highly heterogeneous social and cultural make up of local communities and their environments, indigenous knowledge systems are difficult to 'scale up'. Ranging from minor variations in linguistic terminology to major differences in concepts and beliefs, there is a level of complex diversity among various TEK systems that must be acknowledged. As such, attempts at doing knowledge collaborations at scale must proceed with significant caution, ensuring free prior and informed consent in every step of the process, as there is significant risk involved for indigenous people.

At this stage, the only realistic solution is to facilitate opportunities for 'scaling-up' through relationship-building exercises, shared project activities and/or knowledge exchanges (between indigenous groups, as well as with their non-indigenous partners).

Protecting intellectual property rights of indigenous peoples

The issue of Intellectual Property Rights (IPR) as it relates to TEK has not been given adequate attention especially by the World Intellectual Property Organization (WIPO). Copyrights Commissions need to be strengthened so as to combat the piracy of indigenous knowledge. This is to enable adequate reward of innovative activity, ensure that the society has access to such innovations as well as prevent the unauthorized usage or abuse of such innovative activities. This accentuates the need for bearers of indigenous knowledge to be incorporated into national development planning so as to provide for easier adaptability. Similarly, it will be of prime importance if governments can formulate development policies and programmes which will be channelled through the existent indigenous practices and institutions rather than devising alien ones which cannot easily adapt to the local conditions. It is also important for community members who are sharing certain information to understand how this information can be used and how they can be protected. The relationship between the current Intellectual Property (IP) legal framework and the protection it might afford Traditional Knowledge (TK) and traditional cultural expressions (TCE) is complex.

Traditional forms of IP (patents, trademarks, copyright, industrial designs, geographical indications, trade secrets) provide the rights holders with economic and moral rights over their creations for a fixed period of time. While these traditional forms of IP protection can and should be used by indigenous people to protect their IP rights where appropriate, current Copyright Acts are yet to adequately protect this. This is in spite of discussions both domestically and globally (as discussed earlier in this chapter) to address (among other things) the protection of TEK, to prevent culturally and spiritually offensive uses and to stem the appropriation of culture.

A more integrated approach between municipal legal systems and indigenous laws is clearly required, which allows room for indigenous world views and affords protection of TEKs. Until that development occurs, indigenous communities may want to consider alternatives such as traditional knowledge engagement protocols and private contracts to protect their heritage and promote their culture.

Conclusion

TEK and MSK have been acknowledged as the best of both worlds and may provide the foundational base through which we identify solutions to our most complex environmental problems. One without the other will not accomplish broad-scale environmental protection, and applying only one paradigm can make the environmental situation worse. We have an immense amount of cumulative MSK information and knowledge to do anything, but without wisdom from TEK, it is a lopsided approach – environmental conservation will move nowhere until we fully integrate the two paradigms.

Indigenous peoples' *in situ* knowledge-beliefs-practices have the potential to make significant contributions to meeting contemporary sustainability and conservation challenges globally. However, they are often met with scepticism or simply overlooked as 'traditional' – that is, of the past. Acknowledging and dealing with intellectual baggage that ties many scientists, managers and policymakers to the past is a challenge for knowledge integration efforts.

This chapter has outlined approaches available to indigenous people and their partners to share, use and co-produce the best available knowledge base for decision-making, management and monitoring of biodiversity conservation. Strengthening indigenous conservation institutions and the building of MKS/MEB approach facilitate the weaving of diverse knowledge systems to produce an enriched picture of biodiversity conservation. It describes a means by which indigenous knowledge holders and scientists can collaborate to work with (rather than against) each other's' truth claims by letting each knowledge system speak for itself, within its own context, without assigning one dominant knowledge system with the role of an external validator.

These kinds of intercultural knowledge partnerships form the backbone of collaborative efforts to promote planetary health. The mixed knowledge system is capable of producing transdisciplinary research and monitoring results that harness the strengths of both TEK and MSK and are legitimate, credible, salient and useable in managing and conserving biodiversity. Further, if successful, knowledge integration and co-production exercises can increase stakeholder buy-in and the perceived legitimacy of decisions made and policy formulated. However, implementing the mixed knowledge approach requires coordinated institutional support and sufficient resources to produce useful knowledge that is easily translated into programs of action. To ensure this, there is a fundamental requirement that indigenous peoples, and their knowledge-practices-beliefs, are empowered and have sufficient capacity to collectively organize and mobilize at organizational and institutional levels. It is important to acknowledge the thousands of years of hard work from traditional owners, knowledge holders, practitioners, advocates and researchers.

Notes

1 Aichi Target 18
2 www.ipbes.net/publication/thematic-assessment-pollinators-pollination-and-food-production
3 Discussed in Chapter Three above
4 Discussed in Chapter Two above

References

Agrawal A and Gibson CC, 'Enchantment and Disenchantment: The Role of Community in Natural Resource Conservation' [1993] 27(4) *World Development* 629–649.

Ali MB, *Participatory Mapping as a Tool for Mobilisation of Indigenous and Local Knowledge and Enhanced Ecosystem Governance in Ginderberet, Oroma Region, Ethiopia: A Contribution to the Piloting of the Multiple Evidence Base Approach* (Stockholm Resilience Centre, Stockholm, 2016).

Austin BJ, Vigilante T, Cowell S, Dutton IM, Djanghara D, Mangolomara S, Puermora B, Bundamurra A and Clement Z, 'The Uunguu Monitoring and Evaluation Committee: Intercultural Governance of a Land and Sea Management Programme in the Kimberley, Australia' [2017] 18(2) *Ecological Management & Restoration* 124–133.

Austin BJ, Robinson CJ, Lincoln G, Dobbs RJ, Tingle F, Garnett ST, Mathews D, Oades D, Wiggan A, Bayley S, Edgar J, King T, George K, Mansfield J, Melbourne J, Vigilante T; with the Balanggarra, Bardi Jawi, Dambimangari, Karajarri, Nyul Nyul, Wunambal Gaambera and Yawuru Traditional Owners, *Mobilising Indigenous Knowledge for the Collaborative Management of Kimberley Saltwater Country* (Final Report of project 1.5.1 the Kimberley Indigenous Saltwater Science Project (KISSP); Prepared for the Kimberley Marine Research Program, Western Australian Marine Science Institution, Perth, Western Australia 2018)

Berkes F, *Sacred Ecology* (3rd ed., Routledge, New York, 2008).

Berkes F, 'Community Conserved Areas: Policy Issues in Historic and Contemporary Context' [2009] 2 *Conservation Letters* 19–24.

Berkes F, *Coasts for People: Interdisciplinary Approaches to Coastal and Marine Resource Management* (Routledge, New York, 2015).

Bohensky EL, Butler JRA and Davies J, 'Integrating Indigenous Ecological Knowledge and Science in Natural Resource Management: Perspectives from Australia' [2013] 18(3) *Ecology and Society* 20.

Brondizio E, Settele J, Díaz S and Ngo H (eds),, *'Global Assessment Report on Biodiversity and Ecosystem Services of the Intergovernmental Science-Policy Platform on Biodiversity and Ecosystem Services* (IPBES Secretariat, Bonn, 2019).

Christie M, 'Transdisciplinary Research and Aboriginal Knowledge' [2006] 35 *The Australian Journal of Indigenous Education* 78–89.

Christie M, 'Knowledge Management and Natural Resource Management' in MK Luckert, MB Campbell, JT Gorman and ST Garnett (eds), *Investing in Indigenous Natural Resource Management* (Charles Darwin University Press, Darwin, 2007).

Danielsen F, Burgess ND, Balmford A, Donald PF, Funder M, Jones JP and Child B, 'Local Participation in Natural Resource Monitoring: A Characterization of Approaches' [2009] 23(1) *Conservation Biology* 31–42.

Dixon BA and Wood AP, 'Local Institutions for Management of Wetlands in Ethiopia: Sustainability and State Intervention' in B van Koppen, M Giordano and J Butterworth (eds), *Community-Based Water Law and Water Resource Management Reforms in Developing Countries* (Centre for Agricultural Bioscience International, Oxfordshire, 2007) 130–145.

Donnely-Roark P, Ouedraogo K and Ye X, *Can Local Institutions Reduce Poverty? Rural Decentralization in Burkina Faso* (Policy Research Working Paper, Environmental and Social Development Unit, Africa Region, World Bank, Washington, DC, 2001).

Gorenflo LJ, Romaine S, Mittermeier RA and WalkerPainemilla K, 'Co-Occurrence of Linguistic and Biological Diversity in Biodiversity Hotspots and High Biodiversity Wilderness Areas' [2012] 109(21) *Proceedings of the National Academy of Sciences* 8032–8037.

Hill R, Davies J, Bohnet I, Robinson CJ, Maclean K, Pert PL, 'Collaboration Mobilises Institutions with Scale Dependent Comparative Advantage in Landscape Scale Biodiversity Conservation' [2015] 51 *Environmental Science & Policy* 267–277.

Houde N, 'The Six Faces of Traditional Ecological Knowledge: Challenges and Opportunities for Canadian Co-Management Arrangements' [2007] 12 *Ecology and Society* 17.

International Institute for Sustainable Development, 'Summary of the Second Session of the Plenary of the Intergovernmental Science-Policy Platform on Biodiversity and Ecosystem Services' [2013] 31(13) *Earth Negotiations Bulletin* 1–14.

Johnson, JT, Howitt R, Cajete G, Berkes F, Louis RP and Kliskey A, 'Weaving Indigenous and Sustainability Sciences to Diversify our Methods' [2016] 11(1) *Sustainability Science* 1–11.

Larcom, S., van Gevelt, T., Zabala, A., 'Precolonial Institutions and Deforestation in Africa' [2016] 51 *Land Use Policy* 150–161.

Laurell Å and Åke N, *Plant Species Composition and Plant Uses in Traditionally Protected Forests in Tanzania* (Minor Field Study No. 119, Uppsala, Sweden: Department of Systematic Botany, University of Uppsala, 2000).

Leach M, Robin M and Ian S, *Environmental Entitlements: A Framework for Understanding the Institutional Dynamics of Environmental Change* (Discussion Paper 359; Institute of Development Studies, Brighton, 1997).

Lenjisa D, 'The Significance of Indigenous Knowledge and Institutions in Forest Management a Case of Gera Forest in South-western Ethiopia' [2015] 4(4) *International Journal of Scientific & Engineering Research* 3023–3031.

Maffi L and Woodley E, *Biocultural Diversity Conservation: A Global Sourcebook* (Earthscan, London, 2012).

Martin E, Suharjito D, Darusman D, Sunito S and Winarno B, 'Traditional Institution for Forest Conservation Within a Chnaging Community: Insights from the Case of Upland South Sumatra' [2016] 8(2) *International Journal of Indonesian Society and Culture* 236–249.

Meliyo Joel, Hussein M and Mowo GJ, 'The Baga Watershed Characterization: A Step for Scaling Out Technologies' in T Amede, L German, S Rao, C Opondo and A Stroud (eds), *Integrated Natural Resources Management in Practice: Enabling Communities to Improve Mountain Livelihoods and Landscapes* (African Highlands Initiative, Kampala, 2006) 305–315.

Menzies C and Butler C 'Introduction: Understanding Ecological Knowledge' in C Menzies (ed), *Traditional Ecological Knowledge and Natural Resource Management* (University of Nebraska, Lincoln, NE, 2006).

Mistry J and Berardi A, 'Bridging Indigenous and Scientific Knowledge' [2016] 352 *Science* 1274–1275.

Mowo J, Adimassu Z, Catacutan D, Tanui J, Masuki K and Lyamchai C, 'The Importance of Local Institutions in the Management of Natural Resources in the Highlands of East Africa' [2013] 72(2) *Human Organization* 154–163.

Muller S, 'Co-motion: Making Space to Care for Country' [2014] 54 *Geoforum* 132–141.

Nadasdy P, 'The Politics of TEK: Power and the "Integration" of Knowledge' [1999] 36(1/2) *Arctic Anthropology* 1–18.

Olate R, *Local Institutions, Social Capital and Capabilities: Challenges for Development and Social Intervention in Latin America* (Washington University, Saint Louis, MO, 2003).

Otsuka K and Place F, *Land Tenure and Natural Resource Management: A Comparative Study of Agrarian Communities in Asia and Africa.* (Johns Hopkins University Press, Baltimore, 2002).

Reid WV, Berkes F, Milbanks T and Capistrano D, *Bridging Scales and Knowledge Systems: Concepts and Applications in Ecosystem Assessment* (Island Press, Washington, DC, 2006).

Renwick AR, Robinson CJ, Garnett ST, Leiper I, Possingham HP and Carwardine J, 'Mapping Indigenous Land Management for Threatened Species Conservation: An Australian Casestudy' [2017] 12(3) *Plos One*, 1–17.

Shackleton S, Campbell B, Wollenberg E and Edmunds D, 'Devolution and Community-Based Natural Resource Management: Creating Space for Local People to Participate and Benefit' [2002] 76 *Natural Resource Perspectives* 1–6.

Shyamsundar P, Araral E and Weeraratne S, *Devolution of Resource Rights, Poverty, and Natural Resource Management: A Review* (Environmental Economics Series Paper No. 104, International Bank for Reconstruction and Development/ The World Bank, Washington, DC, 2005).

Sillitoe P, 'The Development of Indigenous Knowledge: A New Applied Anthropology' [1998] 39(2) *Current Anthropology* 223–252.

Tengö M and Malmer P, *Dialogue Workshop on Knowledge for the 21st Century: Indigenous Knowledge, Traditional Knowledge, Science and Connecting Diverse Knowledge Systems* (Workshop Report: Usdub, Guna, Yala, Panama, April 10–13, 2012).

Tengö M, Brondizio ES, Elmqvist T, Malmer P and Spierenburg M, 'Connecting Diverse Knowledge Systems for Enhanced Ecosystem Governance: The Multiple Evidence Base Approach' [2014] 43 *Ambio* 579–591.

Tengö M, Malmer P, Elmqvist T, Brondizio ES and Spierenburg M, *Multiple Evidence Base: A Framework for Connecting Indigenous, Local and Scientific Knowledge Systems* (Stockholm Resielince Centre, Stockholm, 2016).

Tengo M, Hill R, Malmer P, Raymond C, Spierenburg M, Danielsen F, Elmqvist T and Folke C 'Weaving Knowledge Systems in IPBES, CBD and Beyond – Lessons Learned for Sustainability' [2017] 26 *Current Opinions in Environmental Sustainability* 17–25.

United Nations Environment Programme, *Local Biodiversity Outlook* (UNEP, Montreal, 2020a).

United Nations Environment Programme, *Global Biodiversity Outlook* (UNEP, Montreal, 2020b).

Unuigbe, N 'The Significance of the Stewardship Ethic of Indigenous Peoples of Nigeria's Niger Delta Region on Biodiversity Conservation' in C La Follette and C Maser (eds), *Sustainability and the Rights of Nature in Practice* (CRC Press, New York, 2020).

Uphoff N, *Local Institutions and Participation for Sustainable Development* (Gatekeeper Series No. 31; International Institute for Environment and Development, London, 1992).

Verran H (2002) 'A Postcolonial Moment in Science Studies: Alternative Firing Regimes of Environmental Scientists and Aboriginal Landowners' [2002] 32(5–6) *Social Studies of Science* 729762.

Verran H, 'Science and the Dreaming' [2008] 82 *Issues* 23.

Wohling M, 'The Problem of Scale in Indigenous Knowledge: A Perspective from Northern Australia' [2009] 14(1) *Ecology and Society* 1.

World Resources Institute, *Ecosystems and Human Wellbeing: A Framework for Assessment*, ed. A Joseph and E Bennett (Island Press, Washington, DC, 2003) Ch5.

Yeboah-Assiamah E, Muller K and Domfeh K, 'Institutional Assessment in Natural Resource Governance: A Conceptual Overview' [2017] 74 *Forest Policy and Economics* 1–12.

Index

Page numbers in *italics* indicate figures and page numbers in **bold** indicate tables.

Adaptive Environmental Management (AEM) 30–33
adaptive management 29–30
AEM *see* Adaptive Environmental Management (AEM)
Africa: bushmeat consumption in 7–8; indigenous weather forecasting in 39–40; logging in 8; small-scale agricultural systems in 38–39; soil classification in 38–39
agricultural management 38–39, 42–43, 47–48
Akwé: Kon Guidelines 44
Amazon rain forest 1–2, 8
animal agriculture 6–9
Antarctic ice sheets 2
anthrax 9
Articulação Pacari (Brazil) 49
Asia 40–43
Australia 43–44
avian flu 5

Bateson, Gregory 28, 32
bats 3–5, 9–10, 13
Berkes, Fikret 18
Berry, Thomas 21
biodiversity 1, 8, 14n1, 20–21
biodiversity conservation: adaptive management and 29; awareness of 67–68; defining 14n1; health/recreational activities and 67–68; indigenous people and 66; local knowledge systems and 33, 57–64, 66; in Mesoamerica 47–48; small-scale societies and 21; TEK and 18–21, 24, 56, 59; TEK-mainstream science partnerships 67–75; women and 68
biodiversity loss: climate change and 45, **65–66**; deforestation and 5, 8; global challenge of 1; human intrusion and 4; indigenous people and 37, 59; transfer of viruses and 4; zoonotic disease and 64
Brazil 1, 49
bushmeat consumption 7–8
Butler, C. 56

Canada 48
Central Africa 8
China 3–4, 42–43
citizen science 50
climate change: Amazon rain forest die-off and 2; biodiversity loss and 45, **65–66**; deforestation and 3, **65**; expansion of tropical disease and 3; fire management 43; fishing/ocean impacts of **65**; food and agriculture impacts **65**; health impacts **65**; human pandemics and 2–3; ice sheet melting and 2; impact on indigenous lifestyles 3; impact on species **65–66**; mitigation of carbon emissions and 12; species loss and 4; urbanization and 3; weather forecasting and 39–40

Convention on Biological Diversity (CBD) 20, 44, 54–56
coronavirus disease *see* COVID-19 pandemic
COVID-19 pandemic: as environmental tipping-point 11–12; mitigation of carbon emissions and 12; origins of 3–4, 13; zoonotic disease and 4–5, 13, 64
customary rights 57

deforestation: biodiversity loss and 5, 8; climate change and 3, **65**; indigenous people and 64; infectious disease emergence and 4–5, 8–10; livestock farming and 8–9; reduction of 12; tropical 8
dengue fever 3
domestic animals 6, 8–9, 12

East Africa 39–40
Ebola virus 3–5, 9–10, 13
ecological resilience 29
Economics of Ecosystems of Biodiversity, The 20
ecosystems restoration 33, 54
Emerging Infections (IOM) 6
endangered species assessment 48
environmental destruction 1–5, 8–10; *see also* biodiversity loss; deforestation
environmental policy: Adaptive Environmental Management (AEM) and 33; conservation awareness and 68; discourse of 58; gender considerations in 68; government support for 69; health and recreational activities 67–68; holistic picture in 71; indigenous people and 46, 60–61; integrated dialogues and 70–71; knowledge co-production and 70–72; local support for decision-making 69; Multiple Knowledge Systems (MKS) and 60–61; peaceful coexistence with Nature and 12; systemic studies of indigenous knowledge 68; TEK and 20–21
Equator Prize (UNDP) 49
ethics 13–14, 21–24
Ethiopia 39

Europe 44–46
extractive industries 5, 9–10, **65**

Finland 44–46
fire management 43–44
fishing systems 42–43, 45–46, 48, **65**
food-borne disease 11
forest gardening 48
forest management 46–48, 57, 64, **65**
formal institutions 56–57, 68

Gaia Hypothesis 14
Gillespie, Thomas 9–10
global health *see* planetary health
Global North 12
global warming 3, 9; *see also* climate change
Gowtage-Sequeria, S. 6
Greenland ice sheets 2

Hammastunturi Wilderness Area 44
Heiltsuk First Nation 48
HIV 5, 10
human-animal contact: animal agriculture and 6–7; bushmeat consumption and 7–8; domestic animals and 8–9; infectious diseases and 3–5, 9–10, 12; livestock transport and 7; wildlife and 8, 12–13; *see also* zoonotic disease
human pandemics: climate change and 2–3; COVID-19 3–5, 11–13, 64; mitigation plans for 12

ice sheets 2
India 41
Indigenous Knowledge (IK): development and 17–18; environmental change and 45; Intellectual Property Rights (IPR) and 73; knowledge sharing and 66; mainstream scepticism of 63, 74; Multiple Knowledge Systems (MKS) and 60–64, 74; natural resource management and 17–18; rural communities and 18; scaling up 73; social sciences trends and 17; sustainability and 17; systemic studies of 68; types of knowledge in 61–62; weather forecasting

and 39–40; *see also* Traditional Ecological Knowledge (TEK)
indigenous people: biodiversity conservation and 66; biodiversity loss and 37, 59; defining 24n1; endangered species assessment and 48; fire management and 43–44; forest management and 46–48, 64; impact of climate change on 3; impact of English language on 22; Intellectual Property Rights (IPR) and 73; kincentricity and 23; knowledge co-production and 37, 57–61; local knowledge systems and 33–34, 61–64, 66; Respect for Nature 21–23; restorative ethics and 21–23; small-scale agricultural systems and 38–39; sustainability and 49–50; TEK and 18, 20–24, 56; weather forecasting and 39; world views and 20, 23
Indonesia 42
infectious diseases: animal agriculture and 7; COVID-19 3–5, 11–13, 64; deforestation and 5; emerging 5–6; urbanization and 5; wild animal transfer of 3–6, 9–10; *see also* zoonotic disease
influenza virus 4–5
informal institutions 57–58
Intellectual Property Rights (IPR) 73
intergenerational equity 41
Intergovernmental Panel on Climate Change (IPCC) 2
Intergovernmental Science-Policy Platform on Biodiversity and Ecosystem Services (IPBES) 20, 37, 55–56, 61, 64
International Fund for Animal Welfare 13
International Indigenous Forum on Biodiversity (IIFB) 56
International Indigenous Forum on Biodiversity and Ecosystem Services (IIF BES) 56
Intertribal Timber Council 46

Kimmerer, Robin 22
kincentricity 23
knowledge brokers 72
knowledge systems: as beliefs 61–62; best-practice partnership 59–60; as content 61–62; cooperation and 58–59; co-produced 31, 69–71; intercultural brokerage 72; local communities and 33, 57–60; multiple knowledges 60–64, 66, 70–71; as process 61–62; *see also* Indigenous Knowledge (IK); Multiple Knowledge Systems (MKS); Traditional Ecological Knowledge (TEK)

Lassa fever 3
Lee, K. N. 30
Leopold, Aldo 8
livestock farming *see* animal agriculture
local institutions 57–58
local knowledge systems 33–34, 49–50
Locally Managed Marine Areas (LMMA) approach 41
Lyme disease 9

mainstream science (MS): AEM and 32–33; categorization and 28; empirical evidence and 27; hypothesis testing and 28; indigenous scepticism of 63; knowledge sharing and 66; limitations of 23, 28–29; reductive process in 53; scepticism of indigenous knowledge 63, 74; soil classification and 38–39; systemic relationships and 28; weather forecasting and 40; *see also* TEK-mainstream science partnerships
malaria 3
Malawi 38–39
Marburg disease 13
marine resource management 40–41
Maya people 47–48
MEB *see* Multiple Evidence Based (MEB) approaches
Menominee Indian Reservation 46–47
Menominee Restoration Act (1972) 47
Menzies, C. 56
Mesoamerica 47–48
Michael, D. N. 30
Millennium Ecosystem Assessment 20
MKS *see* Multiple Knowledge Systems (MKS)

Index

Multiple Evidence Based (MEB) approaches 60–61
Multiple Knowledge Systems (MKS): data analysis and communication 71–72; Indigenous Knowledge (IK) and 60–64, 74; institutional contexts for 60–61; knowledge brokers and 72; planetary health and 60, *67*
Mustonen, Tero 45

Näätämö River Co-Management Initiative 45–46
native science 53–54; *see also* Traditional Ecological Knowledge (TEK)
natural resource management: adaptive management and 29; AEM and 33; animate role in 57; co-produced knowledge and 31; formal institutions and 56–57; Indigenous Knowledge (IK) and 17–18; informal institutions and 57; local institutions and 57–58; Saami people and 44–45; TEK and 20, 27, 29, 57, 59–60
Nature 12, 21–23, 27
Nipah virus 3, 13
North America 46–48

Pacific Islands 40–41
Patil, Rajan 8–9
pest management 42–43
PETA 13
Philippines 42
planetary health: challenge of 1, 10–11, 64; impact of climate change on **65**; impact of environmental destruction on 2, 8–10; Multiple Knowledge Systems (MKS) and 60, *67*; TEK and *67*; zoonotic disease and 10–11
primates 13
public health *see* planetary health

rainwater harvesting (RWH) 32, 41
Respect for Nature 21–23
rice-fish culture (RFC) 42–43
rural communities 3–4, 18–19

Saami people 44–46
SARS *see* Severe Acute Respiratory Syndrome (SARS)
science, mainstream *see* mainstream science (MS)

sea level rise 2
Severe Acute Respiratory Syndrome (SARS) 3–5, 13
Silko, Leslie Marmon 23
Skolt Sami people 45–46
small-scale societies 20–21
Snowchange Cooperative 45
South America 49
Strategic Plan for Biodiversity 2011–2020 55
Strathcona Regional District (SRD) 48
Summary for Policymakers of the Global Assessment Report on Biodiversity and Ecosystem Services (IPBES) 64
sustainability: forest management and 46; Indigenous Knowledge (IK) and 17; local knowledge systems and 33, 49–50; local participation and 38; marine resource management 40–41; *see also* biodiversity conservation; natural resource management
Sustainable Development Goals (SDGs) 1
swine flu (H1N1) virus 9

Tanzania 39
TEK *see* Traditional Ecological Knowledge (TEK)
TEK-mainstream science partnerships: adaptive environmental management and 31–33; biodiversity conservation and 67, 73–75; conservation awareness and 68; data analysis and communication 71–72; endangered species assessment and 48; fishing systems and 45–46; forest management and 46–47; gender considerations in 68; government support for 69; holistic planning and 71; implementation of identified activities 71; integrated dialogues and 70–71; Intellectual Property Rights (IPR) and 73; intercultural knowledge brokerage 72; local decision-making and 69; multiple knowledges and 69–70; natural resource management and 28–29, 36–38, 50; rainwater harvesting and 41; respectful dialogue and 70; scaling up indigenous knowledge 72; systemic studies of indigenous knowledge 68

Tengö, M. 60
traditional cultural expressions (TCE) 73
Traditional Ecological Knowledge (TEK): AEM and 31–32; animate role of resources in 57; conceptual models for 50; dynamic nature of 19–20; ecological resilience and 29; ecosystems restoration and 54; endangered species assessment in Canada 48; environmental policy and 20–21; erosion of 19; implementation and 71; indigenous people and 18, 20–24, 50; integrated dialogues and 70–71; intergovernmental decisions and 54–56; natural resource management and 20, 27, 29, 57, 59–60; planetary health and *67*; pockets of social-ecological memory and 19; practice and 62; restorative ethics and 21–24; small-scale societies and 20–21; social-ecological resilience and 18–19; systemic relationships and 28–29; unity with natural world and 22–23; verification of 27–28; *see also* Indigenous Knowledge (IK); native science; TEK-mainstream science partnerships
traditional leaders 29, 41, 57, 66, 69
traditional medicinal practices 49
tropical disease 3

Uganda 39
UN Biodiversity Summit 37
UN Decade for Ecosystems Restoration 54
UN Environment Programme (UNEP) 5, 54
UN Food and Agriculture Organization 6, 42
United Nations' Declaration on the Rights of Indigenous Peoples (UNDRIP) 20
United Nations Development Programme (UNDP) 49
UN Permanent Forum on Indigenous Issues 54
Uphoff, N. 58

urbanization: animal agriculture and 7; climate change and 3, **66**; disease containment and 12; infectious disease emergence and 5; planning and development 12; pockets of social-ecological memory and 19; species loss and 9; spread of viruses and 3–4
U.S. Institute of Medicine (IOM) 6

Virchow, R. 5

weather forecasting 39–40
West Nile virus 3, 9
wet markets 4, 10
wilderness area management 44–45; *see also* natural resource management
wildlife: ban on markets for 13; COVID-19 and 4–5; disease transmission between domestic animals and 8–9; habitat destruction and 8–9; impact of climate change on **65–66**; redistribution of 45; trade in 8, 13; transfer of viruses and 3–5, 8–9, 12–13
Wisconsin 46–47
women 68
Woolhouse, M. 6
World Health Organization (WHO) 5–6, 13
World Intellectual Property Organization (WIPO) 73
World Organization for Animal Health 6
World Wide Fund for Nature 41
Wuhan, China 3–4

Zika virus 3, 5
Zoological Society of London 13
zoonotic disease: animal agriculture and 6–7; bats and 13; biodiversity loss and 64; bushmeat consumption and 7–8; COVID-19 and 3–4, 64; global health and 10–11; habitat destruction and 8–10; human fatalities and 10; livestock transport and 7; pandemics and 5–6, 10; primates and 13; terrestrial biodiversity and 8

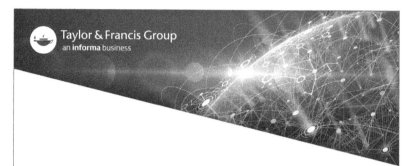